Paterson Great Falls

Paterson GREAT FALLS

FROM LOCAL LANDMARK TO NATIONAL HISTORICAL PARK

MARCIA DENTE

Foreword by Mayor Lawrence F. Kramer

THE History PRESS

Published by The History Press
Charleston, SC 29403
www.historypress.net

Front cover: by Marcia Dente. *Back cover*: image of Alexander Hamilton courtesy of the Library of Congress. *Back cover, bottom*: courtesy of the Paterson Museum.

First published 2012

Manufactured in the United States

ISBN 978.1.60949.725.5

Library of Congress CIP data applied for.

Notice: The information in this book is true and complete to the best of our knowledge. It is offered without guarantee on the part of the author or The History Press. The author and The History Press disclaim all liability in connection with the use of this book.

Contents

Foreword

When Alexander Hamilton convinced George Washington that if this young agrarian nation was to survive and prosper, it must create and develop an industrial capability, he did so at the Great Falls in Paterson.

This vital decision and the events that followed have now been recognized and saluted. The Falls district was designated the birthplace of American industry, and a national park surrounding the district was created.

So all know and appreciate what took place here, Hamilton's and Washington's vision and actions ignited not only the birth of American industrial capability but also massive employment opportunities, economic development, the union movement and, most importantly, the ability to defend ourselves as a nation to meet the challenges of war.

A special salute should always be noted for the men and women of those generations that made it all work at home or on the battlefields. These folks helped meet each challenge when and where we were attacked.

Now from the pen of Marcia Dente, who literally grew up next to the roar and mist of the Great Falls itself and who has lived and loved its history and always understood its national significance, comes her newest book, *Paterson Great Falls: From Local Landmark to National Historical Park*.

An earlier book by Marcia took the reader on a wonderful trip full of events and history that now provides us with the foundation of this fascinating new read.

A most knowledgeable author takes us past its incredible beauty to the serendipity of gifts to our nation. This national park will help prepare a new generation of Americans, especially Patersonians, who will benefit from a deeper understanding of the past, provide a better appreciation of the present and inspire directions for the future.

—Lawrence F. Kramer
former mayor of Paterson, 1967–71 and 1975–79

ACKNOWLEDGEMENTS

This book would never have been possible without the contributions from the following people. My heartfelt gratitude is endless to everyone I've had the privilege of working with on this project and who contributed so much to seeing this book to completion. My last book was written prior to the official designation of the Great Falls as a national historical park. I want to begin by thanking the Paterson Museum, particularly Jack DeStefano, curator, and Joe Costa, photo archivist, for some of the images that appear in this book. They are always accommodating. I also want to thank the American Labor Museum/Botto House National Landmark, specifically Angelica Santomauro, museum director, and Evelyn Hershey, education director, for responding so quickly to my request for images about the Silk Strike in 1913. Once again, a BIG thanks to the Library of Congress, New Jersey Digital Highway, Connecticut State Library Photo Archives, New York Public Library Photo Archives, Texas Rangers Hall of Fame Museum in Waco, Texas, the Municipal Utilities Authority and the Historic American Engineering Record (HAER) for all the images that appear throughout this book.

My deepest thanks go to Steve Dente and Ketrin Basile for giving up the major part of their day and scanning images for this book and putting them on CDs for me. Many thanks to Pat Kramer for his continued support and enthusiasm. When I asked him to write the foreword for this book for me, he agreed without hesitation. With his knowledge, he should be the one writing this book about the Falls. I am deeply indebted to my friend Roberta Papale for her guidance, support and encouragement. Thank you for the time and energy you spent reading through the several drafts and final manuscript, working diligently, making sure the book was ready for publishing.

And lastly, I want to dedicate this book to the memory of my father, Harry, who, with the exception of the four years he served in World War II, lived his entire life on Front Street, just two blocks away from the Great Falls. He has influenced and guided me my entire life and continues to do so. Up until his death in 2007, he always believed in the potential of the Falls and the surrounding area and believed that one day it would be a national park. He would be able to say, with a matter of pride, that he lived near the Great Falls National Historical Park.

All these people made this book possible.

INTRODUCTION

After our forefathers won the War of Independence, Americans fought their second revolution—the fight for economic independence—in the city surrounding the Great Falls. The names of George Washington, Alexander Hamilton and the Marquis de Lafayette are forever tied to Paterson's history.

Before Paterson became the nation's first planned industrial city, the soon-to-be town and its surrounding countryside was called Acquackanonk and was the homeland of the Lenni Lenape Indians. Later the land played host to Dutch settlers, warring Revolutionary armies and a budding Industrial Revolution.

The Industrial Revolution started with the founding of the town of Paterson. Today, both the United States and the world enjoy the fruits borne of seeds planted in the city.

As a leader in the Revolution, first treasurer of the United States and president of the Bank of New York during the end of the eighteenth century, Alexander Hamilton distinguished himself for all time in the nation's history. But his part in the founding of urban Paterson, while nationally significant, holds a very special meaning to Paterson. Hamilton selected the Great Falls (also known as the Passaic Falls and the Totowa Falls) for an ambitious experiment. He promoted the natural power of the Great Falls as an excellent location for textile mills and other manufacture. By the late 1880s, Paterson had become so well known for silk manufacturing that it was called the "Silk City."

Paterson attracted skilled craftsmen, obtained engines from Europe to run the mills and, as a result, produced a large concentration of creative and able people. Paterson made cloth from cotton and silk to clothe a nation and sails to power ships across the world's oceans. Paterson built locomotive engines to unite our nation by rail and aircraft engines to unite our world by air. Thomas Edison installed one of the first hydroelectric power plants in the world using the Great Falls as an energy source. The power plant still provides energy today.

The mills and shops where preceding generations worked are monuments to our heritage. Their homes, great and small, are monuments to their spirit. The mills can be seen in a variety of ways. They can be admired for their handsome architecture or they can be viewed as monuments of America's Industrial Revolution. They are places where people spent a large portion of their lives and were essentially the heart of Paterson's industry. The power and majesty of the Great Falls is a monument to the dedication that makes America both great and humble. As a result, the Great Falls became the nucleus for a burgeoning mill industry. Two manufactures of prominence, the Rogers Locomotive Works and the Colt Gun Mill, where Samuel Colt produced the first Colt 45 revolver, benefited from the water power of the Falls.

The Great Falls became a National Natural Landmark in 1967. In 1971, the Great Falls Preservation and Development Corporation was established to restore and redevelop the historic mill buildings. Had it not been for the intrepid efforts of Mary Ellen Kramer, first lady of Paterson, during her husband Lawrence F. Kramer's three terms of office as mayor, there never would have been a historic district. A major roadway designed to change the delineation of wards and districts in Paterson was strongly advocated in the late 1950s by the state of New Jersey, but Mary Ellen organized and led the battle to block the roadway and preserve what many experts considered the nation's first industrial archaeological site. On June 6, 1976, President Gerald R. Ford visited Paterson and proclaimed the Great Falls/SUM Historic District and its surrounding 119 acres as a National Historic Landmark District. Within this tract are some ninety buildings and a water-power system that uniquely documents the changes in industrial technology from its inception in the early 1800s through the present. This designation culminates years of efforts on the part of many individuals, who felt that the nation's first planned industrial city should remain intact. Former governor James E. McGreevey visited the waterfall, with state and federal officials looking on, on October 21, 2004, and announced the adoption of the Great Falls area as a New Jersey State Park. The legislation signed by President Barack Obama on March 30, 2009, authorized the Paterson Great Falls National Historical Park as part of the Omnibus Public Lands Act of 2009, which included the designation of 35 acres of land. On November 7, 2011, Secretary of the Interior Ken Salazar and Paterson's mayor, Jeffery Jones, with the Great Falls as a backdrop, signed a historic agreement that officially paved the way for the establishment of the nation's 45th national park, 397th unit in the National Park System and the 4th national historical park.

The national park will commemorate the dramatic Great Falls of Paterson and its place in the history of the American Industrial Revolution while preserving its natural beauty and its story to be recounted, again and again, for all those who come to visit the area. Paterson is unique because it is the nation's first planned industrial city and because the historic landmark is virtually intact today.

Chapter One

ACQUACKANONK

The Lenni Lenape and the Dutch

The Passaic River begins its journey in the hilly, wooded region of northern New Jersey. From there it winds and turns through the swampy lowlands along the western edge of the Great Swamp and passes through the gorge of the Great Falls.

The Falls formed approximately thirteen thousand years ago, at the end of the last glacial period. The Great Falls of the Passaic River was originally a geological feature, the basaltic ridge of the Watchung Mountains, through which the Passaic River flowed into an eroded gorge. The Passaic River followed a shorter course through the Watchung Mountains near what is now called Summit, since the previous course the river took was blocked by a newly formed moraine. In its unaltered form, the Falls was a natural site of great beauty. As the ice receded, Glacial Lake Passaic ponded behind the Watchungs; the river managed to escape and found a new and circuitous route around the end of the mountains, and it was there the spectacular seventy-seven-foot Falls was carved out through the underlying basalt.

The gorge and waterfall remained substantially unaltered, and though a bridge and water pipe cross the gorge, the surrounding area has been gradually developed by harnessing the power of the Passaic River. Paterson's roots reach deep into the historic growth of our nation. Long before the colonies united and declared their independence, travelers and visitors flocked to Acquackanonk, the homeland of the Lenni Lenape Indians, to view the breathtaking beauty of the Great Falls. Acquackanonk is said to be derived from three Indian words: "ach-quo-ni-can," meaning brush or brush net; "hanne," meaning a rapid stream; and "onk," meaning a place.

The Passaic River, escaping from a dammed glacial lake, carved a new route through the basalt, ultimately creating the Great Falls. *Courtesy of the author's collection.*

The complete translation is then "a rapid stream where brush nets are set." The name refers to the v-shaped brush nets that were set in the river in shallow places, so constructed that fish became entangled in the twigs. This process yielded bountiful catches during the season when spawning runs of shad and sturgeon were underway. Sturgeon and shad made their way to the foot of the Falls and, because of their size, were especially prized. One of the largest ever reported caught was on August 31, 1817, and weighed 130 pounds. However, with the building of the Dundee Dam at the head of the tidewater in 1858, migratory runs of shad, sturgeon and other species ended. With increasing industrialization along the river and pollution, many species gradually disappeared. Today, sunfish and carp are practically the only fish left. Fishing the Passaic at the Falls is usually unproductive now.

It took about three thousand years for the area to be settled by a nomadic group of hunter-gatherers. The Lenni Lenape Indians knew the Falls as a prime camping and fishing site, which they called Totowa, an Indian word that meant "to sink or be forced down beneath the water's weight"—a tribute to the awesome mass of water. Passaic was a Lenape word, too. Some say it meant "place where the land splits" or "place where the river splits"; others claim it meant "valley." However, the name "Passaic" troubled the early settlers who tried to spell it, creating dozens of variations such as Passaick, Passawicke, Passaya and Fishawack. The early descriptions of the Falls speak of the awe inspired by it and portray in various ways the movement of the water as it tumbles over the rocks and drops to the valley below. Today, they are known as the "Great Falls," for it is indeed an experience to see the mass of water in violent motion and hear it roar as it strikes the rocks below and rushes through the gorge when the Passaic River is at full flood level. Clouds of spray fill the atmosphere, and rainbows appear and disappear as the sun's rays set fire to the ever-changing mists. The Indians never dreamed of harnessing that mighty energy or of compelling it to do their bidding. They left that to the white men.

Since the white man first gazed upon this natural wonder in the seventeenth century, countless visitors have written about it; poets have sung of its beauties; artists have attempted to portray it in line and paint; and it has been photographed from every conceivable angle.

No exact record exists as to who first discovered the Falls. It seems likely that following the settlement of Newark, hunters or land prospectors from that colony wandered up the river and toward the Falls in their exploration of the surrounding wilderness. In all probability, it was known prior to the purchase of Dundee Island at Acquackanonk.

In the 1620s, Dutch settlers, missionaries and trappers came meandering up the river to a spot below the Falls, intrigued by a description given to them

Dutch missionaries and trappers settled near the Great Falls in the 1620s, intrigued by a description given by friendly Indians. *Courtesy of the author's collection.*

by friendly Native Americans. Dutch traders and missionaries began visiting the area near the Falls in the late 1670s. The earliest recorded was Hartman Michielsen, who, in 1679, bought an island in the river from Captehan Peeters, an Indian with an anglicized name. Jaques of Najack, along with seven or eight associates, purchased Acquackanonk from the Indians for $100 to $150 in Dutch guilders. The tract had an abundant supply of wood, enough for building purposes and fuel. The purchase left only one hut with a family of Indians on the tract. Also, in winter 1679, two Dutch missionaries journeyed over the terrain that led these men to the Falls. It took them three hours over the hills, until they came to a high, rocky hill where they heard the noise of water. As they clambered to the top, they saw the Falls below and wrote an awe-filled account of the grandeur of the Falls, what they described as "a sight to be seen in order to observe the power and wonder of God." In their observation, the men found the river making its way to the

The power of the Great Falls attracted visitors to the site long after Alexander Hamilton's picnic lunch during the American Revolution. *Courtesy of the Paterson Museum.*

chasm and saw water pouring through a ten-foot-wide opening with a height they guessed to be about eighty feet. It caused a great roar, and the foam and spray were constantly ascending like smoke and scattering like rain. And in the spray, when the sun shone, the figure of a rainbow was visible. By the end of the seventeenth century, knowledge of this natural wonder had become widespread, and the Falls began to attract numerous visitors.

However, property disputes followed between the Dutch and the Lenni Lenape, and hunting, trapping and the trading of pelts and hides began to deplete the wealth of the Dutch's regional supplies. The Lenape, now exposed to previously unknown European illnesses, were weakened by disease, which curtailed their ability to stem the settlers' further encroachment.

The original settlers of Acquackanonk were plain, hardworking people, tillers of the soil and artisans—weavers, blacksmiths, carpenters, masons or leather workers. Forests were cleared, and trees were felled and hewn into logs to build the first shelters for the pioneer settlers. (Those lofty trees were later replaced by a forest of chimneys from the great mills in Paterson.) The new settlers had no money, so their living was simple and meager. Johannes, Elias and Cornelius Michielsen were said to be the first settlers in Acquackanonk.

The first routes of travel were merely Indian trails—footpaths through the wilderness. Later on, these trails were used by the settlers to lay out their roads. The oldest reference to a public road in Acquackanonk is in a Dutch journal, under an entry dated April 10, 1693, which mentions "the King's Highway" along the river. In 1774, the King's Highway was merely a dirt road that connected Totowa with Acquackanonk Landing (now the city of Passaic).

Settlers were establishing farms in the area, and the Dutch received formal patents, obtained the first tract of land and began farming, the principal occupation at what is now the site of Paterson. In 1682, fourteen Dutch families acquired the Acquackanonk Tract. (When the Dutch settlement was formed, Acquackanonk included parts of present-day Clifton, Paterson and Passaic.) Over time, as the land was passed on to the settlers' descendants in smaller parcels, the economy of the area became more diversified. Some residents became weavers, carpenters and stonemasons; others operated mills, distilleries or stores. In 1693, residents formed a Dutch Reformed Congregation. But the area remained rural and Dutch until the late eighteenth century.

By the turn of the eighteenth century, knowledge of the natural wonders of the Falls had become widespread, and it began to attract visitors from near and far. Many who saw it left a record of their visit by carving their initials or names in the face of the cliff. A favored place for the early graffiti was on

This lithograph represents one of the earliest known views of the city. It was published in 1834. *Courtesy of the author's collection.*

William H. Bartlett was best known for his numerous finely detailed engravings. Between 1835 and 1852, he made four visits to the United States. *Courtesy of the author's collection.*

a large flat rock that became known as the Grotto of Records because of the many initials carved on it. From this spot, there was a beautiful view of the river and the surrounding countryside. General George Washington visited this spot frequently when his army was encamped nearby and conferred with his brother officers. On one occasion, he carved his initials into the trap rock with the year "1778," enclosed in a box with the initials of other officers. These were on the right side of the rock opposite the Grotto of Records and were visible for almost a century.

By 1791, Paterson had become a peaceful hamlet with ten dwellings, a church and an inn called the old Godwin Tavern (also known as the Washington House), which served as a place of entertainment for visitors to the Falls, dating back to about 1750. At the hotel, folios were provided for the benefits of the guests. In these, visitors were encouraged to write their names, the dates of their stays and their comments, just as you would find in a hotel or tourist guest book today. Before the American Revolution, a stage line ran twice a week between Powles Hook (Jersey City) and Godwin Tavern, bringing tourists on a three-day round trip to view the beauty and the power of the Great Falls, second only in size to Niagara. The stage line stated in an advertisement that it was a fine coach that was hung on straps and very comfortable to ride in. There was room for nine passengers, and two more passengers could be accommodated on top of the stage. There was also a rack for trunks and packages in the back. The stage could be seen early Monday and Thursday mornings in front of the Godwin House (better known as the Passaic Hotel), where it waited to take passengers to Powles Hook. The route the stage took as it pulled away from the Passaic Hotel was down King's Highway to Acquackanonk Landing, across Schuyler's Swamp to Powles Hook.

Advertisements in the New York newspapers dating from the 1700s tempted people to take the stage to the Falls. The *New York Gazette and Weekly Mercury*, on September 5, 1774, advertised that Abraham Godwin "has lately built a very commodious house for tavern-keeping about two hundred yards from his late dwelling house, at the foot of the bridge and on the King's Highway," and that he offered his guests "a convenient room for dancing, and a fiddler will always be ready for the service of the ladies and gentlemen who require it. Also a guide to attend any strangers who shall show them all the natural curiosities at the Falls." This advertisement was signed Abraham Godwin, who was often called Gordon.

During the Revolutionary War, Abraham Godwin declared himself with the colonies and was listed as an enemy of the King. Although he owned considerable property and several buildings, he sold it off to Jacob van Winkle, who was well known in this territory as a Tory, in order to satisfy his debts. Godwin purchased another tract of land and built another home for his family.

In 1820, Godwin's son added a long dining room to the rear of the original hotel, and under his management, the Passaic Hotel had a reputation as being the best hotel in the state. But after his death in 1835, the hotel started to decline.

The Godwin House, built just prior to the Revolution, was enlarged, repaired, renovated and altered many times during its long history. This famous and ancient hotel was connected with the area now known as Paterson. In 1915, part of the structure was still standing, but it was condemned and completely demolished shortly thereafter.

During the Revolution, the hotel's proprietor, Jacob van Winkle, wined and dined Washington, Lafayette and other important people. For many years, those who came from distant places to view the wonders of the Great Falls enjoyed its hospitality. After the Revolutionary War, the Godwin family took possession of the hotel again. For many years it operated as a tavern and hotel. General Godwin was always on hand to lead a parade or make a speech for the Fourth of July celebrations. His hotel was the

The first bridge over the Falls was the Clinton Bridge, named for DeWitt Clinton, linking both sides of the chasm. *Courtesy of the New York Public Library Photo Archives.*

place for all meetings of importance in the early days of Paterson. In 1792, the founders of the Society for Establishing Useful Manufactures (SUM) held a meeting there to decide the location of the city of Paterson. DeWitt Clinton held lavish banquets there in 1823, and the Passaic County courts were established there in 1837. The leading citizens of the city entertained distinguished guests in the banquet room. After it was renamed the Passaic Hotel, it played host to Lafayette on his return to Paterson in 1824.

In 1850, William Archdeacon, owner of the Cottage on the Cliff at the Falls, advertised it as a place where target practice ranges were available to military companies and where there was a spacious dining hall attached to the cottage that could accommodate two hundred persons at a time for dinner. Peter Archdeacon, who purchased the grounds around the Falls in 1839, further improved the property and replaced the Clinton Bridge with another covered structure in 1844. The third bridge was built by John Ryle. It was open at the top, with paneled sides, and like the first bridge, it was built of wood. In March 1868, the bridge was declared unsafe and was replaced by an iron bridge with open sides so the Falls would be visible to those who crossed over the chasm.

In 1776, the British won the Battle of Long Island and possession of the Lower Hudson River and Newark Bay. The stage no longer ran, and the isolation of the area at the Great Falls seemed complete except for Indian trappers and shad fishermen. Abraham Godwin went to war and death. He died on February 9, 1777, and was buried at Fishkill, New York. After his death, his wife kept a tavern at her home.

During its long existence as a hotel, there were many different owners in addition to the members of the Godwin family. It was managed by Abraham van Houten, Helmagh van Gieson and Christopher Breeze. On October 1, 1818, Breeze bought the tavern property from Daniel Holsman for $2,750. On April 10, 1822, he advertised it as follows: "For Sale at the house known by the Sign of the Bergen County Hotel on the Hill, At the Great Falls House, farm implements, 2 barn stables. Valuable shad fisheries attached to the farm." He sold the place with other property to Daniel K. Allen on May 1, 1822, for $1,900. The tavern was on Redwood Avenue, between Totowa Avenue and the Falls. When Breeze was about to buy the tavern property on the Falls, Abraham van Houten offered him most of the block bounded by Broadway, Main, Van Houten and Prospect Streets for about the same price as the Falls property. Breeze preferred the latter since Totowa Road was more traveled than Main Street.

Until 1892, it was known as the Passaic Hotel, but in 1893, it lost its classification as a hotel and became a boardinghouse and rooming house. It became a hotel again in 1910 and was called the Continental Hotel, owned by Adolph Smith.

It's been reported that General Washington flew his headquarters' flag over two taverns near the Falls. It was also reported that the Passaic Hotel had a room in it known for many years as the Washington Room. In later years, the room was remodeled as a ladies' retiring room. The commander in chief was supposed to have had a flag over another tavern that was a small, one-story Dutch-type farmhouse, occupying a site on high ground near the Falls known as Monument Heights.

Then from the mines of Pompton and Ringwood, war brought oxcarts and wagons down the Cannon Ball Road with loads of iron missiles for the Continental army. In 1780, some of those forces were assigned to guard the land from the vantage point offered by the cliffs of the Valley of the Rocks, often referred to as the Grand Parade Ground, near the Great Falls. The war also brought the Continental army itself to camp on the Totowa Hills above the wondrous Falls and allowed General Washington to inspect the area. It was during a lull in the fighting that Washington and his staff, Generals Wayne, Knox and the Marquis de Lafayette, decided to enjoy a picnic lunch at the foot of the majestic "Totowa Falls" (another name for Passaic Falls; it is derived from the name Totauwei, "to sink, dive, going under water by pressure or forced under weight of the water"). Totowa is a tract of several thousand acres of land, and the Falls was not called by that name until seventy years after the purchase of the land by the settlers. The Delaware word for waterfall is "sokpehellak," which means "cataract"; another similar word is "sookpehelleu," which means "the water tumbles down from a precipice; for a large or great fall."

Washington's most brilliant aide, Lieutenant Colonel Alexander Hamilton, was also on that picnic and exulted in the pounding of the Falls as they plunged seventy-seven feet into their rocky gorge. It's probable that Washington, Lafayette and others visited the Great Falls and may have discussed Hamilton's vision for the emerging nation. This was Hamilton's first view of the Falls, and it is with Hamilton that the history of the city of Paterson begins.

Chapter Two

First Planned Industrial Center

Alexander Hamilton is arguably the architect of America's economic system, as well as a leading proponent of a unified central government. He was born on a Caribbean island and was the illegitimate child of an absent father and a mother who died when he was only twelve years old. While he was clerking for a trading business, some islanders recognized his potential and took up a collection to send him to New York for a formal education. He arrived in New York City around 1772 or 1773. He went to Kings College, but the outbreak of the Revolutionary War prevented him from graduating. He was appointed a captain of the artillery, became fully engaged in the conflict and, while serving as a trusted and key aide to General George Washington, rose to great prominence.

Alexander Hamilton came to know New Jersey as he participated in the November 1776 retreat from New York into Pennsylvania by way of the Delaware River, the Battles of Trenton and Princeton, the Morristown encampments and the Battle of Monmouth.

Following the Battle of Monmouth on June 28, 1778, Lieutenant Colonel Alexander Hamilton, General George Washington, his aide-de-camp Colonel James McHenry and the Marquis de Lafayette were on their way to the war camp in Paramus. On July 10, they stopped for a picnic lunch at what was known back then as the Totowa Falls.

Colonel James McHenry wrote the only account of the visit to the Falls in his journal: "July 10, 1778, the most momentous day in the history of the Great Falls on our route to Paramus where a part of the army had encamped in order to rest and refresh we had an opportunity of seeing the Falls of the Passaic."

Hamilton's vision for industrial America marshaled a group of private investors to harness the power of the Falls. *Courtesy of the Library of Congress.*

During the stop, McHenry spent a significant amount of time exploring the rock formations around the Falls and described in his journal cracks, crevasses, the basin and river below the Falls. He described the Falls itself as a "place well formed for love." On a particular area he was exploring, he wrote:

> *I was interrupted by a call from the General. When I returned to the encampment I found the General and suite seated under a large spreading oak—"within view of the spray diversified by the beautiful rainbow."*
>
> *The traveling canteens were immediately emptied and a sudden repast spread before us consisting of cold ham, tongue and some biscuits, with the assistance of a little spirits we composed some excellent grog, then chatted away a very cheerful hour—then took our leave of the friendly oak—its refreshing spring—and the meek falls of the Passaic.*

General Washington and Colonel Hamilton enjoyed the picnic lunch at the Falls, and Acquackanonk was favored as a campsite during the Revolutionary War. General Washington sent a group of surveyors to northern New Jersey to investigate possible sites for stationing the Continental troops. The general was already familiar with the area and appreciated it from a military point of view. The Passaic River functioned as a natural barrier, which made Acquackanonk the best defensible spot. Hamilton, viewing the Falls for the first time, was greatly impressed by it as a source of water power. Legend has it that Washington, Hamilton and other staff officers, while having lunch at the Falls that day, "drank Madiera [*sic*] wine and gamboled with the local maidens." Washington's and Hamilton's initials were carved in the rock and are still visible.

In 1780, some of Washington's troops were assigned to guard the land from the vantage point offered by the cliffs above the Valley of the Rocks near the Great Falls. The beauty of the Falls resulted in what may have been an embarrassing moment for one distinguished visitor. The Marquis Francois Jen de Chastellux saw the Falls on his way to visit Washington at his headquarters nearby. Attracted by its roar, he spent so much time admiring its wonders that he was late for his dinner with the general. (An account of the episode is contained in his *Travels in North America*, published in New York in 1827).

Following his military service, Hamilton was a representative to the Continental Congress and vocally advocated for reform of the ineffective Articles of Confederation and the convening of a constitutional convention. Hamilton's thinking was always national in scope. As the nation's first secretary of the treasury, he authored numerous reports that were instrumental in shaping the financial and economic future of the United States, such as the *Report on Public Credit*, *Report on a Plan for the Further Support of Public Credit*, *Report on the Bank*, *Report on Establishing the Mint* and *Report on Manufactures*. Hamilton had been asked by the House of Representatives to investigate the "subject of Manufactures; and particularly the means of promoting such as will tend to render the United States, independent on foreign nations for military and other essential supplies."

Recognizing that America could never be truly free from Britain, or from any other foreign oppressor, as long as our nation remained dependent on foreign goods, Hamilton submitted his *Report on Manufactures* to Congress on December 5, 1791. It is regarded as one of our ablest treatises on government encouragement of manufactory and one of the best treatises on the subject of manufactures ever written. Hamilton set forth multiple arguments in the report on the importance of stimulating American manufacturing. His report recommended an aggressive and active role for the federal government involving the implementation of tariffs, the granting of subsidies, the solicitation of skilled foreign workers and the acquisition and assimilation of technological innovations. The report stated "that a society is forming with a capital which is expected to be extended to at least $1 million on behalf of which measures are already in train for prosecuting on a large scale, the making and printing of cotton goods." The announcement

This close-up view of the Alexander Hamilton statue shows its wear and tear from the elements prior to its restoration. *Photo by Marcia Dente.*

reflected the genius of SUM, which had begun to be organized on the basis of a prospectus published on April 29, 1791, and was soon formalized in an act of incorporation on November 22, 1791. Promoted by Hamilton, the prospectus described a society of investors who would oversee the establishment of a manufactory town focused on one or more factories capable of producing paper, cotton cloth and manufactured products. Initially, Hamilton served as its primary agent. He hired William Hall as superintendent of the society's fabric printing enterprise and Thomas Marshall to superintend the cotton mill. In hiring these individuals, SUM was establishing a deliberate link to the advanced textile manufacturing technology used in England.

In contrast to the beliefs of Thomas Jefferson and others regarding the need to maintain an agrarian society, Hamilton argued that agriculture does not fully employ the workforce available, that industry would help to attract immigrant workers to the fledgling nation and that the diversification of the economy would greatly strengthen the nation's ability to survive and prosper. He also advocated the use of women and child labor and the use of protective tariffs. In his report, he outlined the necessity for the country to move along an industrial course, stressing the importance of a domestic manufacturing capability that would provide articles of prime necessity, particularly cotton goods, and the catalyst for the creation of the Society for Establishing Useful Manufactures. It was considered his most innovative report, providing detailed insight into Hamilton's vision for the United States and its future. Hamilton's *Report on Manufactures* went further than any other report in projecting the economy.

Tench Coxe, tapped as Hamilton's assistant secretary of the treasury in 1790, researched and wrote the first draft of the *Report on Manufactures*, which was revised by Hamilton mainly to add an economic justification for American manufacturing. Coxe was a noted advocate of manufactures and active in a Pennsylvania society for this purpose before his appointment. Hamilton urged Congress to promote manufacturing so that the United States could be independent of foreign nations for military and other essential supplies. In the report, Hamilton wrote, "Not only the wealthy but the independence and security of a country, appear to be materially connected with the prosperity of manufacturers." In addition to national independence, manufacturing would provide a path to equality in the global market. To Hamilton, the absence of substantial manufacturing in the United States was a gaping hole of opportunity that desperately needed to be filled. The report was submitted in 1791 at a time when Hamilton's political enemies were particularly active. Because of their opposition (primarily that of Thomas Jefferson, who was the secretary of state, James Madison and other members of the Republican Party) and because of the far-reaching implications of the plan, Congress was not enthusiastic and failed to act on it. Many prominent citizens were also skeptical of the young nation's ability to raise capital and begin manufacturing at a sizable scale. Shortly before issuing the report, Hamilton had joined in supporting Coxe's plan for a manufacturing society operated by private interests that enjoyed the support of government.

Coxe and Hamilton also collaborated on the formation of SUM, which both foresaw as an object lesson on American manufacturing. Coxe can take some credit for originating the idea for SUM, based on his desire to put into use a patent he had acquired in March 1791 for an improvement in the Arkwright spinning machine (the first machine in the world to successfully spin many strands of cotton yarn without the need of a skilled operator).

At about the same time, he presented to Hamilton a plan to create a joint stock company to finance a cotton factory through a loan backed by government securities, but Coxe's proposed method of publicly financing SUM was not adopted. In 1792, Coxe wrote a strongly worded rebuttal to the insinuation that public monies had been invested in SUM and subsequently lost through the speculation of SUM's governor, William Duer.

A prospectus for SUM was drawn up and published on April 29, 1791, as a collaborative effort by Alexander Hamilton and Tench Coxe. The prospectus expounded on Hamilton's arguments for manufacturing more finished products by corporations, even using public subsidy if necessary.

Alexander Hamilton's statue looks out toward the Great Falls, gazing upon their power and dramatic beauty, envisioning a great manufacturing center. *Photo by Marcia Dente.*

It called for the establishment of a manufacturing town supported by private investments and devoted to the society's manufactures, who produce a multitude of different products, from linens to paper to cotton cloth to beer. While no specific site was mentioned, Hamilton viewed New Jersey as the logical place for the venture. The report emphasizes the practicability of extensively manufacturing cotton in the United States, and he had viewed New Jersey as the one logical place in mind for his "New National Manufactory."

In the latter part of the eighteenth century, a number of Hamilton's associates supported creating large industrial establishments in America. The Revolutionary War had established the political independence of the nation, and the aim of Hamilton and his cohorts was to establish its independence in commerce and manufacturing. In searching for an industrial capital, Hamilton remembered the town where he had stopped for lunch and selected the land surrounding the Great Falls and the powerful Passaic River for the proposed manufacturing settlement.

A letter from William Hall to Hamilton dated September 1791 most likely helped Hamilton in his consideration of the Falls area. In his letter, Hall wrote,

> *Sir, Last night Mr. Mort and myself returned from the Passaic Falls—one of the finest situations in the world (we believe) can be made there. The quality of the water is good and in sufficient quantity to supply works of almost any extent, everything necessary as to situation is here to be found... The situation so far exceeds our expectations that we are very desirous you should see it.*

The site was the land adjacent to the Great Falls. The Falls' surrounding terrain, in addition to the seventy-seven-foot drop into a deep chasm below, had all the characteristics that were essential for the utilization of water power at that time.

As a sort of supplement to his plan for manufactures, Hamilton and his former treasury department assistant, William Duer, founded the Society for Establishing Useful Manufactures, New Jersey's first corporation. Supported

Paterson was named after Governor William Paterson by the Society for Establishing Useful Manufactures in appreciation of his efforts to promote the society. *Courtesy of the Library of Congress.*

Once the Alexander Hamilton statue was cleaned and his long-missing cane was replaced, city officials rededicated the statue on May 26, 2004. *Photo by Marcia Dente.*

by private investors, it was an eighteenth-century industrial park to be built in Paterson, New Jersey, complete with the snappy fiscal acronym SUM. SUM acquired the necessary land and constructed America's first water-power system to attract entrepreneurs and workers who could incubate new inventions. SUM fulfilled the vision of its founder for more than 150 years, from George Washington's first term as president through the end of World War II.

Hamilton was ahead of his time. He believed that hard work and learning, rather than social status or inherited wealth, mattered most in the new nation. He was a courageous early opponent of slavery and a strong advocate of immigration. His plan for economic independence for the new American republic at the end of the eighteenth century called for tapping the inestimable power of the Great Falls. It had what he needed, and he visualized a potential source of power in its waterfalls and behind the great cataract.

This statue was created by Franklin Simmons in 1905 and was once placed in front of city hall and then the post office. It was moved in 1964 to its current location. *Photo by Marcia Dente.*

True to Hamilton's vision, manufacturing and innovation spread rapidly throughout the city, and people of any class, creed or color soon found the ability to climb higher than they had ever imagined. Paterson became a city where a former slave, a poor farmer or a recent immigrant could build a better life. In Paterson, Hamilton launched what we have come to call the American Dream.

Chapter Three

Society for Establishing Useful Manufactures

The Society for Establishing Useful Manufactures was the motivating force behind much of the early development of Paterson. From its inception in the late eighteenth century, SUM was keenly interested in understanding the energy needs of manufacturers, perhaps never more so than from the 1890s to 1910, when the society worked to end the mills' use of water from the raceway and switch them over to its central steam and electric power plants.

The Great Falls of Paterson, which grew as an industrial town throughout the 1800s, provided water power to SUM. The philosophy of the society was to organize a center for the production of goods for sale. Under the society's long-term management of the Falls (over 150 years), the industrialization of the area passed through three great waves, centered first on cotton, then steel and finally silk.

In an appeal for the establishment of manufactures in the United States, issued in early 1791, the prospectus of the society stressed the importance of an independent American economy. Therefore, SUM was organized, initiated and founded by Alexander Hamilton to harness the water power at the Great Falls site. The authorized capital of the new company was $500,000, which was an enormous sum in those days.

SUM convened in New Brunswick for its first meeting in August 1791. Directors were selected and included the following: William Duer, John Dewhurst, Elias Boudinot, Alexander Macomb, Royal Flint, Benjamin Walker, Nicolas Low, John Bayard, John Nelson, Archibald Mercer, Thomas Lowring, George Lewis and More Furmans. Several were from New York, and six from New Jersey. Most were financiers, and the board lacked experienced membership in actual manufacturing.

Many possible mill seats in the mid-Atlantic region were suggested for SUM, but localities in New Jersey were most frequently recommended, and the location of present-day Paterson seems to have been the front-runner from the start. New Jersey was clearly favored because of its proximity to New York City and Philadelphia, the young nation's two largest port cities and population centers. The availability of a reliable and abundant water source at the site was the primary criterion. It was understood by the time of SUM's founding that water-driven machines were the future of the industry. The inherent potential of the Great Falls was immediately apparent to anyone who visited New Jersey's foremost natural wonder. The Great Falls captivated the imagination with a drama none of its competitors could match.

On November 22, 1791, New Jersey governor William Paterson signed the charter (which Hamilton is believed to have been heavily involved in drafting) under which the society was established. The charter granted to the society by the New Jersey state legislature allowed the grantees freedom to designate the locality of operations, time of beginning or period of duration of their operators without restrictions. In addition to manufacturing, the society was authorized to improve navigation by constructing canals, lakes and dams, as well as to levy tolls and raise $100,000 by lottery. Property was exempt from state tax for ten years and from local, municipal and county taxes forever. The charter also stated that the inhabitants within six miles might at any time be incorporated and form a municipal government with the usual powers, privileges and duties.

With the founding of SUM, Paterson became the first industrial city in the world that was centered on a major source of power used for a variety of industries. Hamilton had a vision for the area surrounding the Passaic Falls that included factories, mills and shipping facilities that would supply the state and the region with both raw materials and manufactured goods for sale and trade. His intention was to turn the area surrounding the Falls into what he called a "National Manufactory" for the new republic.

A city grew out of the seven hundred acres SUM held above and below the Great Falls, and the first citizens were primarily workers at local factories. William Duer, a New York capitalist and prominent businessman of the time, had been an assistant to Hamilton at the treasury, prior to Tench Coxe. He was chosen as the first governor of SUM, and Archibald Mercer was chosen as deputy. Duer was raised and educated in England and moved to New York as a young man in 1768. He was known for a friendly disposition and eloquence that aided in his successes. During the Revolutionary War, he served as a deputy adjutant general for the New York troops and also on the New York Committee of Correspondence. He became a delegate from New York to the Continental Congress and later was appointed to the Board of War. He was particularly known to be prone to speculative ventures and was a key figure in the corrupt Scioto Corporation, an infamous group of land speculators in Ohio from 1789 to 1792.

Large-scale manufacture of cotton goods was the principal purpose for selecting an area for a manufacturing center. The next order of business was to designate a site, and after many meetings, the final selection was the Great Falls. The quarries, abundance of timber (needed for fuel) and building materials were the winning factors.

The recently elected governor of SUM proposed to tap the waters of the Passaic River above the Falls by means of a great canal that would extend for nearly seven miles before carrying water back into the Passaic River at its head of tide at Acquackanonk. The next order of business was deciding the town name, which was actually decided upon before the site was selected. Some favored naming it Hamilton because of Alexander Hamilton's contribution in establishing SUM and for selecting the site. Others favored Paterson in honor of William Paterson, the governor, who granted the charter. With Paterson's support, the state assembly and council quickly awarded SUM the liberal charter conveying exceptional powers.

On May 17, 1792, SUM convened to officially authorize the purchase of seven hundred acres of land adjacent to the Falls and dispatched a group of directors to purchase the land. The name Paterson was officially adopted. The committee appointed to lay out the new town was authorized to borrow $70,000 on the stock of the society and decided that water from the river above the Falls should be conducted to a location selected for erecting a cotton mill, buildings and houses for the workmen. The seven hundred acres of land and the river above the Falls were purchased for $8,230.

SUM bought land above and below the Falls to ensure complete control over its water-power potential. Its first priority was to put into place the infrastructure necessary to provide water power for the vast enterprise. The original plan to construct canals from above the Falls and have them empty into the river below proved too costly. SUM then embarked on a short-term program to construct a cotton mill, a weaving operation, an establishment for printing calicoes, a sawmill and housing for workers.

The motives of Hamilton and those of Duer and his associates were different. Duer was driven by speculation, and Hamilton was interested in demonstrating the value of industry in the growth of the nation. SUM directors were also more narrowly focused, reflecting the smaller-scale operations in which they had experience. Another problem facing the society was the lack of technology and skilled workers. Both the directors and Hamilton agreed that the best way to get manufacturing underway was to bring skilled English workers to Paterson and build the same modern equipment they were using in Britain. The goal was to get manufacturing up and running as soon as possible.

Initially Nehemiah Hubbard, a prominent merchant of Middletown, Connecticut, was elected to the positions of engineer and superintendent of SUM on January 21, 1792. It was unclear if he was aware that he was a candidate. On May 3, he declined the position, but Hamilton wrote a letter to Hubbard persuading him to reverse his decision. Hamilton's pleas were to no avail. Hubbard wasn't relocating to New Jersey to assume the position, so he put forward Pierre Charles L'Enfant's name as a candidate to fill the position. (L'Enfant was the temperamental and extravagant engineer who had recently laid out the plans of Washington, D.C.) The positions offered to Hubbard and L'Enfant were not quite the same. Hubbard had been offered

Pierre L'Enfant was hired by Alexander Hamilton to design and construct the raceway system to harness the power of the Great Falls. *Courtesy of the Library of Congress.*

the position of "Superintendent-General of the Works of the Society" and L'Enfant, "Agent for Superintending the Erection of the Works." L'Enfant was hired with the understanding that his tenure would not exceed one year. He was a planner and engineer, and he was not to oversee the operations of SUM works following construction. Hamilton chose L'Enfant as being one of a few professional engineers who might be able to solve the design of the water-power system. Despite friendly relations with Hamilton, L'Enfant proved to be a problematic choice. He came to Paterson a disgruntled man, and he would leave Paterson even more discontented.

His recommendation included a plan for conveying the Passaic waters three to four miles to Great Notch before using them. He also suggested running streets and avenues in Paterson two hundred feet wide at right angles, regardless of the rocky and hilly terrain. Major L'Enfant was an enthusiast who saw in Paterson the possibilities of an industrial capital. His city plan for Paterson, however, was never carried out, and any actual drawings he may have illustrated are lost. His ideas were nevertheless too grand for the men who had put up the capital for Paterson, although he did design water-powered raceways that would ultimately be modified for use in the city.

From what is known of L'Enfant's training and previous experience, there is little to lead one to believe that he possessed the credentials to design a complex and sophisticated water-power system such as Paterson required. The great age of American canal building had yet to truly dawn, but there were few others in the starting nation who had any experience at all. Hundreds of mill raceways had been constructed throughout North America by this time, but none compared in complexity with the system that SUM was proposing to build.

Financial panic set the young nation back between the years 1792 and 1793. The panic was largely caused by the massive amount of speculation, much of it by William Duer. SUM was significantly affected because Duer and other directors had taken or invested SUM funds elsewhere. Duer had unsuccessfully brokered a scheme to corner the market in United States government securities and the takeover of the Bank of New York. The effects were instant, and a number of the original investors left. The subsequent financial embarrassment that resulted from indiscretions taken with federal funds and some speculative ventures soured Duer's relationship with Hamilton and threatened the economic security of the nation. Subscribers were now unwilling or unable to invest, and SUM lost its momentum. Hamilton took direct action on behalf of SUM and sought a loan from the Bank of New York. Duer was thrown into debtors' prison in New York, where he subsequently died. The nation's financial system, meanwhile, was thrown into chaos.

Financially crippled, the remaining directors of SUM turned to Hamilton for guidance. He volunteered his leadership until a new superintendent could be hired. Hamilton suggested that the society push forward immediately with its plans for the cotton and weaving mills but recommended that any other planned industrial pursuit be put on hold for the time being.

Hamilton unofficially served as both the manager and governor of the Paterson site. The task of immediate recovery was enormous. Despite the obstacles, Hamilton continued to be dedicated to his grand manufacturing experiment. He attended board meetings and visited Paterson despite a bout with yellow fever.

Philip Schuyler, Hamilton's father-in-law, was instrumental in some of the decisions that followed. The owner of several mills, he was familiar with recent advances in power technology. Hamilton asked for his advice on the factory's location and the layout of the water-power system. Schuyler visited the site on May 29, 1792, and the two of them determined that the most prudent location for the factory was at the Great Falls. Lands were purchased under the rights of the organization's charter, which had been signed the previous year by New Jersey's governor, William Paterson.

In the broadest of terms, Schuyler was responsible for the basic outline of the water-power system that would eventually be constructed; Hamilton served SUM in a senior advisory capacity. He didn't hold a directorship but was present at the meetings when crucial decisions were to be made. On August 1, 1792, Hamilton was selected to serve as the society's agent and to prepare a plan for both the town of Paterson and the area extending from above the Falls to the site of SUM works. After a preliminary examination for the site selected for Paterson, L'Enfant gave a written report to the directors of SUM on August 19. He discussed the merits of several scenarios for transporting the water to the proposed site of the society's works. However, L'Enfant's proposals were too extravagant and had to be abandoned, as his plans called for a relatively complex and lengthy water-power system. In the fall of 1792, L'Enfant pressed ahead with the construction projects under his own charge while Hamilton and the society's directors were still looking for a superintendent of works.

On February 18, 1793, the position was offered to Peter Colt, shipping merchant and treasurer of the state of Connecticut, and he accepted. Colt arrived in Paterson and soon found he did not like the situation. Much needed building had not yet been done, workers were unhappy and, to his estimation, progress was behind. Plus, L'Enfant was nowhere to be found. Colt was hesitant to proceed without L'Enfant's plans, but L'Enfant had gone to Philadelphia to prepare a more detailed version of his plans and to also negotiate a new assignment—that of a design for a house for a Philadelphia financier. When he returned to the Paterson site, the directors held a March meeting and proposed the construction of a canal system on the same lines as had originally been suggested by William Duer, as well as the purchase of all the necessary lands. But L'Enfant would not bow to this directive and left for Trenton, deserting Colt and the project. L'Enfant left Paterson in 1793 and never returned.

Colt, though untrained as an engineer, proceeded as best as he was able. He was now in charge of all SUM construction activities. He continued construction of the industrial buildings as finances would allow. Colt was a practical, capable and resourceful man who was able to successfully complete the construction of the society's cotton mill and water-power system where the Frenchman's ambitious scheme had failed. The Bull Mill was the first to be constructed; it was a small-frame cotton mill powered by an ox. Colt's plan was simple, quick and the most cost effective. The mill exemplified Colt's direct and obvious approach in satisfying the need to supply the society with a cotton mill with water power. He was charged with overseeing the design, construction, operation and maintenance of the water-power system and was responsible for planning and coordinating the construction of the original mill complex and housing for employees.

The result was a system that worked. A canal was completed in January 1794, and water power became available later that year. In June 1794, the second cotton mill was constructed of stone and wood. Peter Colt did his best to see the plans through to fruition and continued to serve in the capacity of supervisor; yet despite Colt's improved management, the enterprise continued to decline. In 1796, at an emergency meeting, failing finances finally forced SUM directors to abandon their efforts and cease operations. The society dismissed the majority of directors from their duty only five years after the signing of the charter. Colt served the society until it suspended operation in May 1797 and then resigned his position and left Paterson for Rome, New York.

With the abandonment of L'Enfant's plans, the priority shifted from providing a source of water power for a range of mill sites to a source of water power for just one, the cotton mill. Paterson's water-power system could be, and eventually was, expanded.

In its early years, SUM had somewhat of a turbulent existence. The society spent too much of its scarce capital on building factories and importing expensive machinery from England. By 1795, SUM was in the red, and the stockholders abandoned it. The society relapsed to a paper existence, and Paterson's population sank from five hundred to forty-three. By 1796, the society's efforts to build its own mills had failed, mostly because of the slow profits it generated. They were not enough to cover its start-up costs. In January 1796, SUM closed its only mill and ceased manufacturing. Its activities were confined to the management of its extensive real estate in Paterson, including the water-power right. Hamilton's envisioned manufacturing enterprise then entered a lengthy period of land leasing and water-power development that enriched other aspiring industrialists. It would never live up to the charge of its far-reaching charter to deal and trade in its own manufactures.

Following its acquisition by Roswell Colt, SUM transformed into a real estate company that leased mill sites and water rights. Colt more or less reshaped SUM into a revenue stream for himself and his family. He even took advantage of a long-forgotten clause in the charter that allowed it to operate a lottery and raised upward of $100,000 in capital funds. At about the same time he was reinvigorating the society, Colt was also working to develop an image of himself as a financial baron and began casting himself as the leading citizen of Paterson. Since Paterson was still largely a company town administered under SUM's original charter, Colt, as the chief stockholder and governor of the organization, wielded a remarkable degree of power and served as the primary focus of civic authority in the community until the formal founding of the township of Paterson in 1831.

Although now a real estate venture rather than a manufacturing colossus, SUM ultimately prospered. In 1800, part of the cotton mill was being used, and a few other manufacturers rented out mill sites, breathing a small bit of life into the all-but-abandoned site. Despite a fire that destroyed the cotton mill, a new raceway was cut in 1807, paid for by surplus income from the leasing of the mills. This was the first large investment made in Paterson in over ten years. The availability of additional power allowed for two more cotton mills to be built. Higher domestic demands for textiles came with the War of 1812, and the city began to grow and prosper. At the close of the war, the market became flooded with foreign goods, and Paterson endured its second setback. Mills became idle and workers were dismissed during this period. With the completion of a second canal in 1829 and the expanding of available water power, the city began the process of renewal again.

By 1815, thirteen water-powered cotton mills were operating beside the Falls and were operated by over two thousand workers. As a result, the population of Paterson grew from five hundred in the 1790s to over five thousand by 1820.

In both 1834 and 1837, a third setback for Paterson occurred when banks failed due to massive speculation. Paterson's industrial future would be realized when industry began to pick up. It wasn't the success of SUM as Hamilton had envisioned it, but the realization of manufacturing diversity and use of an immigrant workforce would occur in Paterson and last into the next century.

Chapter Four
PATERSON'S RACEWAY SYSTEM

SUM's water-power system was built in phases over a period from 1792 to 1846. The society's founders and engineer Pierre L'Enfant had a complex water-power system in mind, but there were differences of opinion as to the plan and its technical details. In the first phase of development, financial and managerial difficulties severely limited construction and required scaling back before the evolving plan could be more than partially executed. In 1796, as a result of that experiment, SUM nearly went bankrupt, and construction stopped. In the early nineteenth century, activities resumed, and SUM took on the role of water power and real estate developer.

SUM's raceways are man-made channels with an elevation of approximately twenty-two feet between each level. The arteries of the water-power system are the three canals known as the upper, middle and lower raceways. The middle raceway is the oldest of the canals, dating from the mid-1790s. The immediate purpose of the middle raceway was to supply the Yellow Mill (which was later expanded and known as the Essex Mill). It eventually served the Colt Gun Mill (circa 1836) and, before that, a rolling mill (circa 1811–13). The years 1800–11 saw the expansion of the raceway system as an implementation of Pierre L'Enfant's and Peter Colt's ideas.

Harnessing the energy of moving water wasn't a new idea. It dates back to water wheels used in ancient Greece, and Paterson is the oldest American community integrating water power, industrial development and urban planning. It was the abundance of inexpensive energy provided by the Great Falls' raceway and water-power system that attracted countless creative and innovative men and women who built and continually improved industries

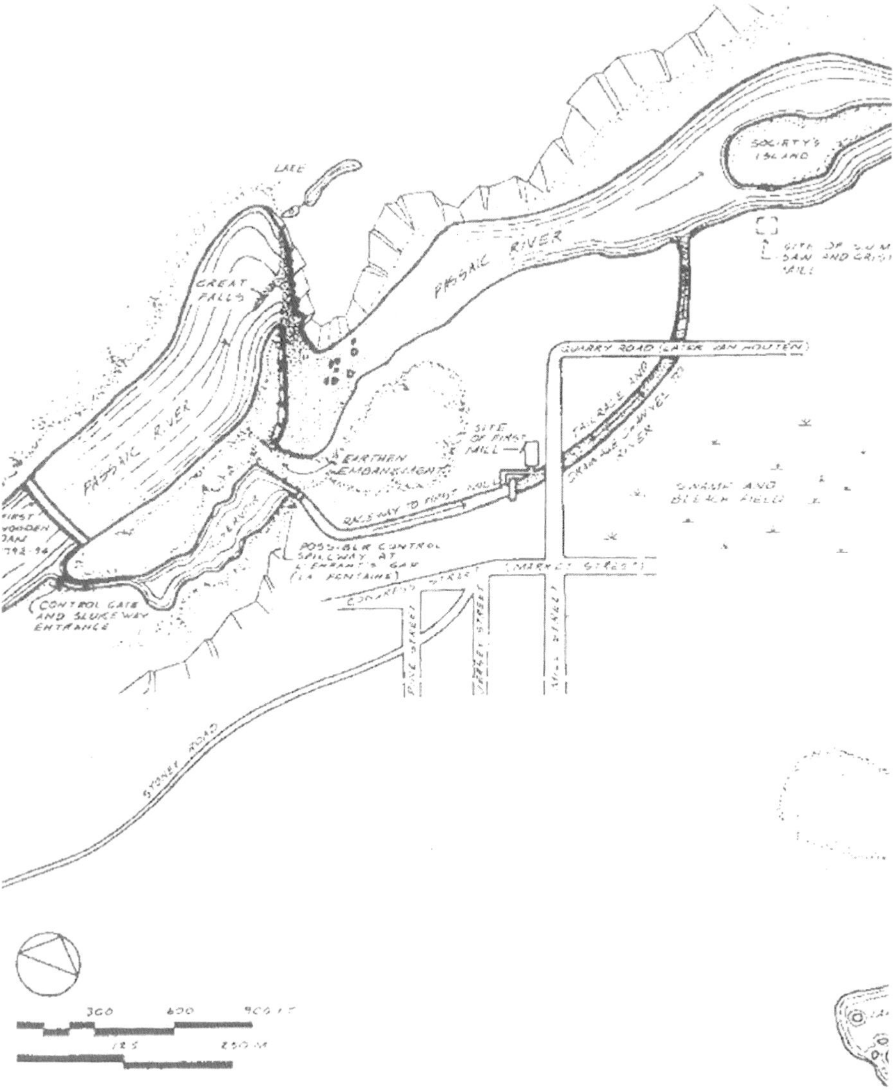

This depicts the first raceway plan, originally designed by Pierre L'Enfant and modified by Peter Colt, with water being diverted by a wooden dam. *Courtesy of HAER.*

such as textiles, textile machinery, the Colt revolver, Rogers's and other steam locomotives, silk manufacturing, flax and jute production, early Wright aircraft engines and apparel manufacturing

A major reason for the Great Falls' designation as a National Historic Landmark was the early harnessing of its water-power resources. Hundreds of mill raceways had been constructed throughout North America by this date, but none compared to the complexity of the system SUM was preparing to build. There were at least four stages of development of the Great Falls Historic District's water-power system. The first, between 1792 and 1794, provided for the basic water supply system and a portion of the middle basin. Between 1800 and 1807, the lower raceway along Boudinot Street (Van Houten Street today) was added. Additions that were made between 1827 and 1846 were considered the most extensive and are what mainly form the system as it exists today.

This massive hydraulic project included damming a major river and sending its power through a system of canals and raceways more than a mile long. It stands as a unique monument to American industrial ingenuity still today, representing the earliest planned engineering project in the United States using water power. Paterson did not have a government until 1831; SUM owned and ran everything. When Paterson finally obtained a town charter, only minimal powers were granted to the people, and SUM maintained its tax-free status. One of New Jersey's first strikes occurred in Paterson in 1794, when workers asked for schools for their children. They were ignored. The Colt family insisted on having streets named after them but declined to contribute to their cost, and Jacob Rogers, son of the

The first wooden dam was built to divert water from the Passaic River into the raceway, passing through to the first SUM mill. *Courtesy of the Library of Congress.*

locomotive tycoon Thomas Rogers, refused to donate a small tract of land for a city hospital, saying he didn't owe anything to Paterson.

The first plan for diverting the waters of the Passaic for powering the mills was drawn up by Pierre C. L'Enfant, and it included a transportation canal over part of the watercourse and aqueduct, as well as the construction of a reservoir to ensure a supply to the mills in periods of low river flow. The costly plans (and L'Enfant's lack of desire to stay within SUM's financial means) were abandoned, and L'Enfant was replaced by Peter Colt.

Construction of the raceway system began in 1792, and by 1793, the raceways were almost finished. Fifteen houses had been built, each twenty-four by eighteen feet, and each included a cellar, garret and a quarter acre of ground. The houses rented for $12.50 yearly or sold at the price of $250, payable within twenty-nine years.

Colt used aspects of L'Enfant's work, and in mid-January 1794, a channel from the river and floodgates was completed. The early raceway system operated from 1794 to 1799 and drew water from the Passaic River through a wooden dam above the Falls called L'Enfant's Gap, near the bend in the river. The ravine was converted into a reservoir, out of which the current passed into what is now the middle raceway. The water then entered into a reservoir and passed through the raceway to a flume and water wheel to provide power for operating a mill. After leaving the mill, the water flowed back into the Passaic River through a drainage channel. It ran 150 yards through this stream to the society's first factory. Shops for printing, bleaching and dyeing had been established by June 1794, and the canal was finished the same year and placed into operation.

When SUM originally planned the town of Paterson, it selected textiles as its chief industry. This industry evolved from the small shops, which did carding and fulling. During the months that followed, considerable yarns and fabrics were made. Bleached and unbleached muslin goods bought in New York were on hand, ready to be made into calico prints and other materials. The society manufactured little and, in January 1796, closed its only mill. Leased by John Park in 1800, the mill was used for spinning candlewick and coarse yarns but was destroyed by fire in 1807 and never rebuilt. This was around the time the society's activities were confined to the management and occasional improvement of its extensive estate in Paterson.

Even though SUM found it necessary to discharge its numerous employees, sell its raw materials and other personal property and discontinue its manufacturing entirely, it did not abandon its waterworks and other improvements. In the first decade of the 1800s, business activity at the Great

The expansion of the raceway system began in 1800 with the extension of the raceway to supply the mills with water. *Courtesy of HAER.*

Falls began to improve, and plans were made to extend the canal. They constructed a second canal in 1807 beyond the first mill site in order to supply the Essex Mill. Headraces and tailraces (the latter being a canal that rids the system of water once it has been used by the mill) were constructed west of

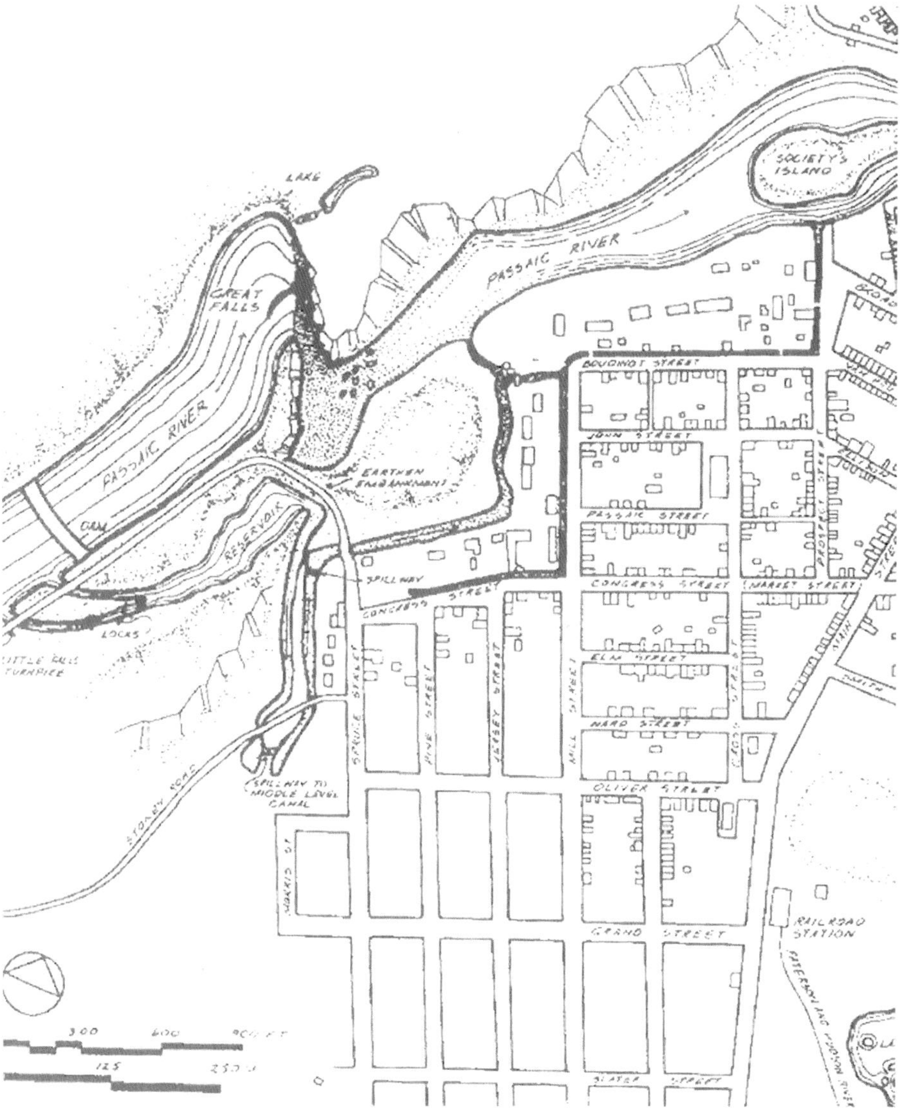

This major modification in the raceway system began in 1827, due to the lack of water for the additional mill sites. *Courtesy of HAER.*

Mill Street and still exist today. This improvement added about five hundred feet of mill lots along the street and increased the depth and capacity of the middle raceway. Additional improvements were made to allow a second tier of mill sites using water at the elevation of the tailrace from the middle canal as the headrace for the new sites. These were located between the river and today's Van Houten Street. Water from the canal went through each lot and returned to the river via individual tailraces. A spillway at the east end of Boudinot Street handled excess water. Each of the above two improvements had a head of twenty-two feet available.

With mill activity growing rapidly in the late 1820s, SUM undertook its first major realignment of the raceway and power system in order to provide water for a new upper tier of mill sites. This third expansion of the system was the most elaborate and expensive. In 1827, the lower raceway was built along Boudinot Street (Van Houten Street). It was the addition of a new upper tier of mill lots that required that the level of the whole system be raised almost to the base of the river to gain the aforementioned twenty-two feet for the new sites. The dam at the end of the ravine was raised and enlarged. The deep gap was enlarged and partially filled to raise the water level, and after passing through, the water made an immediate ninety-degree bend along the face of the ridge for almost five thousand feet. The

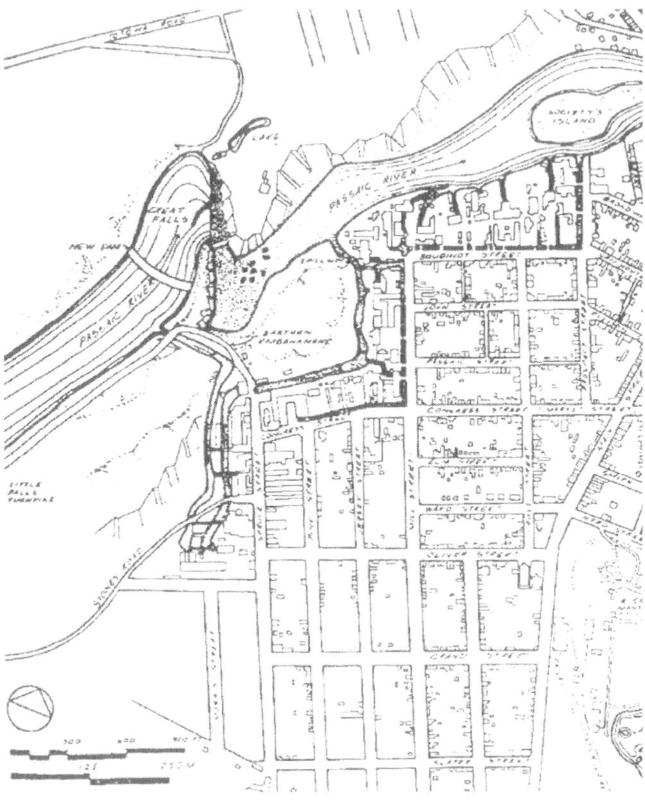

The final alignment of the raceway forced the engineer to abandon the reservoir and channels from the river due to leakage through an embankment. *Courtesy of HAER.*

new canal was cut into the hillside with an embankment to hold the water. Water for the middle canal passed through the upper canal and the new tier of mill lots. This required the tailrace for the new upper group of mills to be higher than the old middle canal. Tailraces on Spruce Street were raised on an embankment ten to fifteen feet high. SUM was forced to sell water rights to newcomers, contingent upon an adequate supply to other mills.

Finally, in 1838, the last modification to the Great Falls Raceway and Power System was made. Leakage through the earthen dam around the reservoir forced the SUM engineer to abandon and then fill in the reservoir and channels from the river. A new channel was cut into the rocky river ledge, a masonry dam was built downstream to divert the river into the new raceway and water flowed through a gully on top of the earthen embankment (which had earlier been a dam for the reservoir) and into the upper raceway. Throughout the nineteenth century, the Great Falls Raceway and Power System was the primary source for manufacturing. After 1850, many of Paterson's new mills were located outside the Great Falls Historic District and used steam as a power source instead of turbines powered by water.

A new masonry dam was built downstream and turned the river into the raceway through a new channel cut into the rocky river edge. *Courtesy of HAER.*

Use of the water through the raceway system continued to expand in the twentieth century as industries prospered in Paterson. However, by 1900, it was apparent that the individual water wheel using water from the raceway was generally less efficient than a central station generating hydroelectricity from the river. If the entire machinery of a mill was not turning, then the full horsepower of the water wheel driving the machinery was not necessary. As the flow of water was decreased, the efficiency fell markedly. Similarly, unless all three of the raceway levels were using precisely the same quantity of water, there would be an imbalance between them and water would be wasted, supplying the one with the heaviest demand while others used only part of the water.

The three-tiered raceway system remains substantially unaltered in its appearance, though SUM added two feet to its dam in 1864 and a concrete sluiceway to supplement the SUM gatehouse at the river entrance. This system has been used continuously since the 1790s and is a landmark of American mechanical and civil engineering heritage.

On June 23, 1982, New Jersey governor Thomas H. Kean and Paterson mayor Lawrence F. Kramer celebrated the opening of Raceway Park, a

The SUM gatehouse, at the entrance of the raceway, stands in the same place it did in 1846, although it no longer functions. *Courtesy of the Library of Congress.*

reconstruction of the water-power system that cost $2 million. Mary Ellen Kramer cracked a champagne bottle against a historic wheelhouse wall, signaling the release of a twenty-two-foot-high fall of water over a historic spillway and into an equally historic water raceway below. Two tiers of raceways, a restored nineteenth-century wheelhouse, two wooden bridges and a gatehouse were all part of the four-acre Raceway Park, built with Green Acre Funds.

"Reconstructing this ingenious eighteenth-century water-power system was like putting pieces of history back together again," said Mayor Kramer, describing the wooded area located near the Falls that was redesigned as a recreational and educational resource.

Chapter Five

The Great Falls Power Plant

Before the SUM power plant, there was only the swamp, a huge waterfall and a handful of Dutch and English settlers. As the population of Paterson grew, demands on the Passaic River water supply increased, and in early 1910, SUM decided to convert the mills along its raceways to electric power due to chronic water shortages that began to affect Paterson's mills. The time had come to terminate the long-standing practice of leasing water from the raceways for industrial power. SUM had begun to consider the feasibility of building a hydroelectric plant at the base of the Great Falls. It was determined that a modest-size hydroelectric plant could produce three times the power of the individual mill power system using half the water. Water levels sometimes fell so low that SUM was unable to provide water to the mills. To meet this challenge, SUM joined in agreement with other users to better manage its water supply. Eventually, SUM fell under control of a trust company that decided that the water used to power Paterson's mills would be more profitably sold as potable water. Therefore, from 1912 to 1914, SUM cut through the cliff near the Falls and built a hydroelectric plant to use the water of the Passaic River, supplementing the function of the raceway and increasing the power available. The raceway system still functions as a source of industrial process water and serves as a standby fire protection reservoir. Over a mile of raceways spreads out though the Great Falls' SUM area.

The hydroelectric-powered generating plant built between 1912 and 1914 served to end the mill's reliance on water from the raceway. SUM also built an auxiliary coal-fueled steam plant that supplied steam for processing in

This 1913 photograph shows the Passaic River bed upstream from the Great Falls with a walkway running across the river bed. *Courtesy of New Jersey Digital Highway Archives.*

the mills through a system of underground pipes. Additionally, it supplied electric power when the river ran low, but the plant was demolished in 1960. The concrete walls and foundations of Overlook Park are the remains of the steam power plant.

Water is drawn into the hydroelectric plant from the river in a channel called the forebay, cut into the hard rock of the cliff. Gates control the water's flow, and screens capture debris and trash before it enters steel tubes called penstocks. The power plant houses the turbines and generators for producing electricity and is connected to the upper-level forebay and gatehouse by an eighty-five-foot stair tower. The gatehouse houses equipment related to the monitoring and transmission of electricity from the power plant, as well as work and storage space for operations related to the gates and trash racks.

Designed by the Thomas Edison Electric Company, the hydroelectric plant produced 4,849 kilowatts and operated until 1969. In 1945, SUM sold the Great Falls power plant and all of its property to the City of Paterson, which leased the plant to Public Service Electric & Gas (PSE&G). In 1969, the plant was shut down due to age, modest size and the need for upgrades. Since oil was cheap at the time, the city neglected its valuable power plant and let it deteriorate. During the 1970s, a group of citizens interested in preservation looked into the possibility of returning the plant to operation. With the oil crisis in the 1970s, the idea of renovating the power plant moved forward

The Great Falls Hydroelectric Plant under construction in 1913. This photograph shows the reinforced concrete foundation, which can still be seen today. *Courtesy of the Library of Congress.*

This photograph shows the progress of the hydroelectric plant, which called for a 4,849-kilowatt facility to be constructed. The plans were drawn up by Thomas Edison Electric Company. *Courtesy of the Paterson Museum.*

despite turbulent economic times. Through a combination of federal grants and private investments, they came up with a plan.

The dates 1791 and 1914 and acronym SUM appear above the door to the power plant—the first for the establishment of the society and the second

This photograph is of the original doors to the hydroelectric plant. The doors were kept intact during the restoration of the building in the 1980s. *Courtesy of the Library of Congress.*

for the year SUM built the plant. The hydroelectric plant had a maximum capacity of 6,500 horsepower, with its four huge turbines generating 21 million kilowatts per year.

Thomas Edison Electric Company drew up plans for a 4,849-kilowatt hydroelectric facility, which operated from 1914 until 1969. The powerhouse was declared a National Historic Landmark in 1971. In 1984, the plant was restored with three new handmade replacement turbines added out of the four, leaving the fourth as a nonoperating original still in its historic place. The Great Falls Hydroelectric Company and PSE&G entered a power purchase agreement on September 6, 1985. In 1986, the plant was restarted and generated 11,000 kilowatts per hour, enough electricity to supply eleven thousand homes. Recently, the plant produced nearly $400,000 worth of electricity in four months, which it sold to PSE&G. The agreement was amended on February 9, 1998, and PSE&G continues to purchase all electrical energy from the facility.

This photograph shows the Great Falls Hydroelectric Plant shortly after it opened in 1914, making more efficient use of water by turning it into electric energy. *Courtesy of the Paterson Museum.*

On June 29, 1981, the Municipal Utilities Authority (MUA) was established by ordinance, with five commissioners appointed by the mayor and with the charge of reactivating the hydroelectric plant. On June 7, 1982, the MUA acquired the hydroelectric plant and approximately ten acres of surrounding land on both sides of the Great Falls. With the assistance of a $1,330,000 grant from the Department of Energy, both MUA and the City of Paterson prepared renovation plans and secured an operating license from the Federal Energy Regulatory Commission. But rather than risk the city's bonding capacity, MUA solicited proposals from private energy developers to invest private capital and develop and operate the plant under a long-term lease.

The MUA selected Synergics, Inc., of Annapolis, Maryland, which created a limited partnership, the Great Falls Hydroelectric Company, to renovate and operate the plant. It invested $14,500,000 in the work, which was done in accordance with historic standards. On June 29, 1987, with the statue of Alexander Hamilton overlooking the scene, along with dozens of schoolchildren and other spectators, government officials rededicated the refurbished hydroelectric generating plant to help ease the nation's dependence on foreign oil. "This is a very historic site," said John Topalian,

The Great Falls Hydroelectric Station was rededicated on June 29, 1987. The restoration was hailed as a success. *Photo by Marcia Dente.*

project manager for the Great Falls hydroelectric station. "This is where the nation's economic base got started." The old turbines and generators from 1914 were replaced with more efficient modern units. One of the original units was left in place for historic value. (During this time period, I worked for John Topalian as his administrative assistant and was involved with the restoration process.)

The renovated dormant plant went online in December 1986 and generated up to eleven thousand kilowatts, an increase of 125 percent over the original equipment, producing one-third of the Paterson area's electrical needs.

The Great Falls again powers Paterson and the surrounding area. The hydroelectric generating facility consists of a number of components, including a dam, headrace canal (called a penstock), intake structure, turbines, generators, draft tube and tailrace canal.

Since the renovated plant was rehabilitated and recommissioned, there have been three transfers of ownership of the limited partnership. The plant

When the hydroelectric plant was restored in the 1980s, three of the four original turbines and the generators were replaced, each with a capacity of 10,950 kilowatts. *Courtesy of the Municipal Utilities Authority.*

is now operated by Algonquin Power and Utilities Corporation of Toronto, Ontario, which began ownership in September 2000 and sells all electric power generated by the power plant to local New Jersey utility PSE&G. One man single-handedly operates and maintains the plant. He was brought on board from day one when Algonquin agreed to run the plant for the MUA.

Water supply in the Passaic River is not constant and continues to be a challenge at the Great Falls. In late winter and spring, water is usually plentiful, and the plant may operate at full capacity. In the late summer or during droughts, the river sometimes runs so low that the plant doesn't have enough water to operate and shuts down. The plant competes with other users of water to fill reservoirs to supply homes and businesses. A minimum of water is always allowed to spill over the Falls for scenic pleasure and to maintain environmental quality for fish and other aquatic life living downstream. The plant also has a fish bypass, an escape route for any fish that become trapped in the forebay.

Once the hydroelectric facility went online, the new equipment produced 125 percent more electricity. A private developer from Toronto, Canada, operates the facility. *Courtesy of the Municipal Utilities Authority.*

The picturesque steel-arch bridge over the chasm was built in 1888 by the Passaic Water Company to carry a water pipe and for pedestrians to use. From 1849 until after 1913, the company pumped water from just above the Falls and distributed it to homes and businesses throughout Paterson. The stone walls visible in low water at the head of the chasm are the remains of the water company's dam. In 1983, the arch bridge was closed to pedestrians, and a second water pipe was added by the Passaic Valley Water Commission to the deck of the arch bridge, replacing the pipe that was previously carried by an adjacent suspension bridge. As a result, a new pedestrian bridge was substituted.

The power plant is more than a building. It's an entire system for capturing the energy in falling water and then converting it into electricity. The power plant is situated at the base of the Great Falls to take advantage of the natural geography. Water is drawn from the river above the Falls and enters buried steel tubes called penstocks. The penstocks, which are

This interior view of the hydroelectric plant shows the four original turbines in the plant in 1914. *Courtesy of the Library of Congress.*

angled down at about forty-five degrees, direct the rushing water to one of the three active turbines. The turbines look like propellers or pinwheels. The water rapidly spins the turbines, the turbines spin the shafts and the shafts turn the generators. The generators transform the mechanical energy using properties of electromagnetism. A wound copper wire called a coil spins between the poles of magnets, creating a high-voltage electric current.

Hydroelectricity is "green" energy because it does not require burning of fossil fuels, like oil or coal. The power plant produces no significant waste material. The water is returned to the river as clean as it was when it entered the plant.

Chapter Six
Growth of Industry Around the Falls

After the Revolution, we had four million commodity-hungry Americans, and Paterson became the prototype for all planned industrial cities in the world. It became the second-largest cotton-spinning city in the United States and provider of cotton duck sails for the navy, the nation's leading producer of locomotives and linen thread, the greatest manufacturer of silk in the entire world and a center for the fabrication of steel. Additionally, it was where the submarine and the Colt revolver were invented, the producer of Wright airplane engines and a training ground for many of America's machinists.

The sounds of the machines were deafening. The air was hot and humid and filled with lint. The windows had to be nailed shut. This was the life of the men, women and children who produced cotton, silk and locomotives, working ten-hour days six days a week. Still, thousands of immigrants came to Paterson to find work. The immigrants worked hard and held high hopes for the future. They learned skills, and the more successful ones moved to new parts of town. They developed strong family and neighborhood ties and emphasized the importance of providing better opportunities for their children. Soon, the labor force organized and demanded better working conditions, and though it was a long and sometimes bloody struggle, the workers won the battle for shorter hours and higher wages. The people who worked in the mills were proud men and women, and their dedication to hard work shaped American labor.

Since textile manufacturing did not flourish under SUM's management, the society, in 1796, ceased its textile operations but continued to control the rights to the water of the Falls and began renting space in its mills to independent manufacturers. Growth was slow at first, but the British embargo of 1807 and the War of 1812 led to a renewed emphasis on American manufacturing, and more entrepreneurs opened mills around the Falls. Unfortunately, the war was followed by a depression, and many of these mills closed. But by the 1820s, the Paterson economy was booming again, and the number of mills doubled by the early 1830s. The mills attracted many immigrant workers, increasing Paterson's population to nine thousand by 1832.

In 1802, at Alexander Hamilton's's insistence, the firm Kinsey, Crane and Fairchild built a paper mill. It began production and became the first paper mill in the world to produce paper in rolls instead of separate sheets. Charles Kinsey, the senior member of the firm, was the inventor of the process, but his partners insisted on abandoning the manufacture of paper for the manufacture of cotton.

Nestled at the base of the upper raceway, the Ivanhoe Mill Wheelhouse, constructed in 1865, provided power to the Ivanhoe Paper Mill complex. The Ivanhoe Paper Mill produced 7,500 pounds of paper a day. Henry Butler, son of a paper mill owner, came to Paterson in 1837 and began paper manufacturing in the Passaic Mill. In 1850, he constructed the Ivanhoe Mill and continued his papermaking enterprise as the Ivanhoe Manufacturing Company, which became one of the most popular brands in the nation. The Ivanhoe Mill produced tissue and writing paper, and Butler obtained several government contracts to produce paper for currency. (Paper money has an unusual texture, as it contains silk fibers.) Originally a ten-building complex, the Ivanhoe Mill's only remaining structure is its wheelhouse, located between the upper and middle raceways. Alfred Gibbs Campbell, an African American industrialist and newspaper publisher, remains a little-known figure associated with the mill. He lived most of his life in Paterson, where he spent thirty-five years as manager and later as part owner of the mills. In addition to his work at the mills, Campbell achieved significant wealth earlier on from the manufacture of patent medicines and perfumes. In 1886, the Ivanhoe Mill went into the first of a series of bankruptcies. The Ivanhoe's wheelhouse was restored in 1981, minus the water wheel and turbine. The round brickwork on the side of the wheelhouse facing the spillway shows where the flume that brought water from the upper raceway into the water wheel ran. Today, a group known as the Ivanhoe Artists Mosaic meet and hold events at the Ivanhoe.

By 1812, Paterson had at least half a dozen cotton-spinning plants among mills. Samuel Colt had a rolling mill for sheet iron and nail factory with seventy-eight employees who turned out seven tons of nails a week. There were plants making wire and brass, weaving plants and plants that manufactured army uniforms. The cotton plant closed, however, in 1815, with the end of the War of 1812. The price of American cotton subsequently rose, leading to consumer preference for cheaper British cotton.

Samuel Colt

Samuel Colt, the man who would popularize the revolver in America and remake the industrial landscape, was born on July 19, 1814, in Hartford, Connecticut. Colt took a circuitous route to founding Colt's Patent Firearms Manufacturing Company. He loved science, even as a boy, and while working as an indentured worker on a Glastonbury farm, he read an encyclopedia on science rather than the Bible. The stage for his later success was further set when he worked in a textile mill that his father owned in Ware, Massachusetts.

A post–Civil War slogan proclaimed, "Abe Lincoln may have freed all men, but Samuel Colt made them equal." This would have been music to Colt's ears had he lived long enough to hear it. Yet, even before his death

Samuel Colt challenged his mechanical ability and inventiveness in 1836 to create a revolver that was later said to have "won the West." *Courtesy of the Library of Congress.*

at the age of forty-seven, he knew that the invention of a weapon capable of firing without reloading was a tremendous success the world over. Some nineteenth-century historians have gone as far as to say that Samuel Colt's invention altered the course of history. But when all was said and done, no man could deny that Colt had achieved a level of fame and fortune known to few other inventors.

As a direct result of his invention and the marketing sales success that followed, Sam Colt and his firearms played a prominent role in early American history. So popular was the Colt revolver during the latter half of the 1800s that it was perhaps the best-known firearm not only in the United States but also in Canada, Mexico and many European countries. To this day, the name Colt suggests firearms to most Americans.

Colt's success story began with the issuance of a U.S. patent in 1836 for the Colt firearm, equipped with a revolving cylinder containing five or six bullets. Colt's revolver provided its user with greatly increased firepower. Prior to this invention, only one and two-barrel flintlock pistols were available. The revolving cylinder concept is said to have occurred to Sam Colt while he was serving in the navy, aboard the ship *Corvo*. There, he observed a similar principle in the workings of the ship's capstan. During his leisure hours, Sam carved a wooden representation of his idea. The principle was remarkable in its simplicity and its applicability to both longarms and sidearms. But Colt's idea was not an instant success. Many people at the time preferred the traditional flintlock musket or pistol to such a novel weapon.

It should be noted that Samuel Colt did not invent the revolver pistol—he perfected it. Elisha Collier designed a revolving pistol in 1818. Colt first saw an Elisha Collier revolver while in the Far East. It was a weapon that used a flint and steel-firing device to set off the powder consecutively in the several chambers of the gun's magazine. Colt realized that a mercury cap and nipple system, both newly developed mechanisms, could be adapted to Collier's early revolver. By 1830, many single-shot flintlock guns were being converted to the newly invented firing method. Colt returned to America in 1832 and paid Boston and Baltimore gunsmiths to build a number of test revolvers using cap and nipple designs as the firing mechanisms on the guns. To pay for the work and support himself, Colt traveled around the eastern states and gave demonstrations on the effects of "laughing gas" (nitrous oxide). When inhaled, the strange gas caused people to laugh involuntarily and, if inhaled in quantity, made dentistry painless.

On March 5, 1836, Colt built his first plant in Paterson, New Jersey, then one of the country's fastest-growing manufacturing centers. Colt's uncle and successful local businessman, Roswell Colt, helped him market his guns by introducing him to investors in Paterson and New York City. The Colt Mill in Paterson was a multistory structure built near the Great Falls. A weather vane in the shape of a gun sat on top of a bell tower, and the picket fence that surrounded the building had pistols for its tops. Together, they formed the Patent Firearms Manufacturing Company with capital stock listed at $300,000. Sam Colt, however, only owned a single $1,000 share and was not an officer or director, but the new corporation hired him as one of its salesmen. He held a contract to sell the Patent Firearms revolving weapons, drew on a Patent Firearms expense account and was paid a commission on every weapon that he sold. At age twenty-two, he became the firm's chief salesman and new business promoter. Sam went off to England, where, in 1836, he obtained a patent protection in England and then returned to America to obtain a patent protection.

In 1836, Colt invented his famous Colt Paterson revolver, later known as "the gun that opened the West." He soon developed three different models: the pocket, the belt and holster and two types of long armor rifles. In all cases, gunpowder and bullets were loaded into a revolving cylinder while the

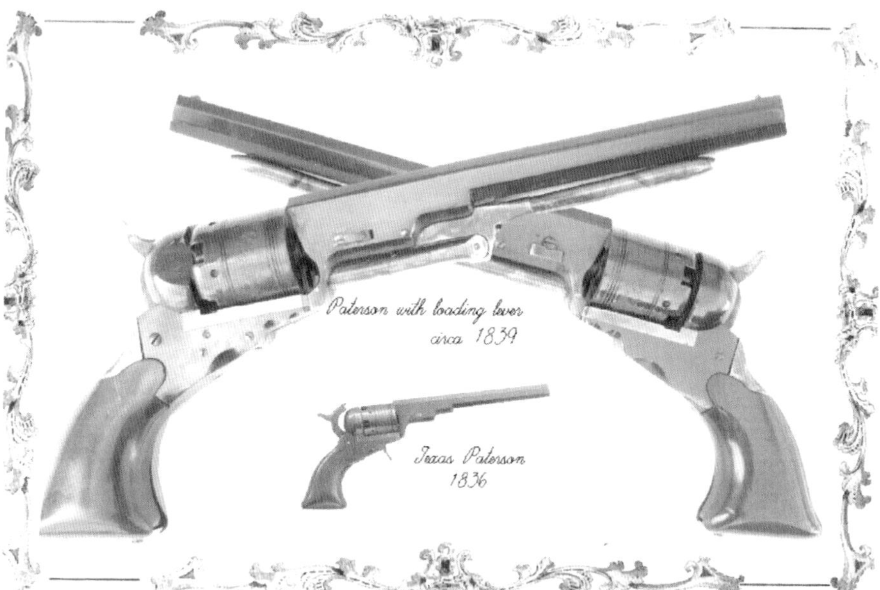

These are two of the models of the Paterson Colt revolvers that were made by the Patent Arms Company before it failed in 1842. *Courtesy of the Texas Rangers Hall of Fame and Museum, Waco, Texas.*

primer was placed into a nipple located on the outside of the cylinder, where it would be struck by the hammer when the trigger was pulled.

Colt demonstrated his weapon on the steps of the Capitol Building in Washington, D.C. Unfortunately, when the gun was fired, it startled the horses tied to President Van Buren's carriage. The horses bolted, destroying the rig and killing its driver. Because of that, Colt's gun was labeled useless and unreliable. Despite the generally favorable performance of the product in the hands of early buyers, sales were sluggish, and though the U.S. government purchased small quantities, those quantities never seemed to exceed one hundred.

By 1840, the Patent Firearms Manufacturing Company was on its last leg. By this time, Colt was heavily indebted to the company because of several loans granted to him to cover his expenses and for guns used by Colt in his attempts at promoting their sale. Ultimately, in 1842 a dispute with John Ehlers led to a court case between Ehlers and the Patent Firearms Manufacturing Company. The Chancery Court of New Jersey ruled against the company, and the assets of the firm were exposed to public sale. Colt was forced to close and began bankruptcy proceedings. He even sold his wooden models. Colt then turned his attention to selling the U.S. government on the ideas for waterproof ammunition, underwater mines for harbor defense and, in association with the inventor Samuel Morse, the telegraph.

Within a few years, Colt reentered the field of gun manufacture. Many of the critical lessons learned during his less than successful sojourn in Paterson served him well in his future ventures. Samuel Colt's actual involvement with Paterson lasted only a few years (1836–41), and during this period, he was rarely a resident in the city. But the name of the charismatic, famous inventor and salesman has been irreversibly imprinted on the popular history of Paterson and has largely overshadowed the civic memory of the other Colts who made far greater contributions to the development of the city.

During 1845, certain units of the U.S. dragoon forces and Texas Rangers engaged in fighting the Indians in Texas. They credited Colt's firearms for their success in defeating Indian forces. U.S. War Department officials were favorably impressed. When the Mexican War began in 1846, Captain Samuel H. Walker of the U.S. Army traveled east and looked up Colt. The two collaborated on the design of a new, more powerful revolver. Within a week, the U.S. Ordnance Department had ordered one thousand of the newly designed revolvers, which Colt called the "Walker." Colt suddenly found himself back in the firearms business but without a gun mill. His friend Eli Whitney Jr., son of the famous inventor of the cotton gin, had large facilities in Connecticut and

On March 12, 1863, Paterson was startled by a fire alarm. The Gun Mill was ablaze in a matter of minutes, and the building was reduced to ruins. *Courtesy of the Library of Congress.*

gave Sam use of part of his mill. The Colt revolver moved to Connecticut, where the U.S. order was completed and shipped by mid-1847. In the same year, Colt borrowed money from his banker cousin Elisha Colt and other Hartford businessmen to lease a factory in Hartford, where he adapted the system of interchangeable parts to the mass production of the guns. A patent renewal in 1849 gave Colt a domestic monopoly on revolver development until the mid-1850s. In 1851, Colt bought property in Hartford's South Meadows, where he built the Colt's Armory, completed in August 1855. The armory was topped with a blue onion-shaped dome, on which stood a bronze rampant colt. Sam Colt died unexpectedly on January 10, 1862, at the age of forty-seven, having produced more than 400,000 weapons in his lifetime. His estate was reportedly worth $15 million, an enormous sum for the time and tantamount to more than $300 million today. Control of the company remained in the hands of Colt's widow and her family after his death.

A fire in February 1864 destroyed the Hartford factory and adjacent office structure, suspending all but certain military production for almost three years.

The Patent Firearms Manufacturing Company was Colt's first attempt at mass production, but the revolvers were unsuitable for military service. It closed in 1842. *Courtesy of the Library of Congress.*

Colt's widow ordered the factory rebuilt exactly as it was and to be as fireproof as possible. Construction was completed in 1867, the year the factory began producing Dr. R.J. Gatling's machine gun, a semiautomatic, using a hand-operated crank to turn a cluster of six to ten barrels while feeding ammunition into the breech. In 1901, Mrs. Colt sold the company to Armstrong & Schimer, a New York financial house. In all the years that have followed, more than thirty million revolvers, pistols and rifles bearing the Colt name have been produced, almost all of them in plants located in the Hartford area.

Morris Canal and Thomas Rogers's Locomotive Industry

In spite of its manufacturing strength, Paterson was geographically isolated. The tons of cloth and iron produced in the town were transported by horse, wagon and sometimes oxcart. The 1830s ended this insularity with the opening of the Morris Canal and the beginnings of what was to become a great railroading center. In the beginning, however, SUM was involved in a dispute with the Morris Canal and Banking Company, as the company had placed a dam on the Rockaway River and diverted water for its own purposes. This reduced the volume of flow over the Falls and threatened SUM's ventures. A subsequent court decision allowed the Morris Canal to be built without disrupting the water supply to the Falls.

The Morris Canal crossed New Jersey and connected Newark Bay and the Atlantic with the Delaware River. It also gave Paterson direct access to the anthracite of Pennsylvania. The Morris Canal was conceived as an industrial cargo route that carried freight, usually coal and iron ore from the mines, but was also equally popular for recreational use. Canalling was enjoyed by entire families, and those who didn't own a boat could pay the fare and ride between Newark and Paterson. Canoes were used on the canal too. Swimming was another pleasure, as was ice-skating in the winter. No one used the canal on Sundays since it was a day of prayer. The canal, however, was never a financial success, and the real culprit in its demise was the steam locomotive.

The ingenuity of a young machinist gave Paterson its first railroad. Thomas Rogers was born in Groton, Connecticut, in 1792 but moved to Paterson in 1812. Trained in carpentry and then as a blacksmith, Rogers formed businesses in Paterson designing and building machinery for textile manufacturing. In 1832, he teamed up with two New York City financiers, Morris Ketchum and Jasper Grosvenor, to form the manufacturing firm of Rogers, Ketchum and Grosvenor. The company diversified, making, among other items, small parts for the newly developing railroad industry.

The production of railroad locomotives and rails in the United States followed earlier developments in England. Colonel John Stevens of Hoboken constructed a steam wagon in his yard in 1825. In 1829, Peter Cooper of New York built the locomotive Tom Thumb and placed it into service on the newly constructed Baltimore & Ohio Railroad.

Up until 1832, all locomotives were made in England, and the first train for the new, short railroad line between Paterson and Jersey City was ordered from there. The locomotive was sent in boxes to a machine shop along one of Paterson's raceways. The owner of the shop, Thomas Rogers, who was a machinist, put the locomotive together but not before he made a pattern of every piece. In less than a year, he built a locomotive, which then went on an excursion toward New Brunswick via Jersey City. Thereafter, he manufactured "iron horses," making the necessary adaptations to build engines to suit the roads. (The English had changed roads to suit the engines, a more expensive process.) Rogers also put the body loosely on wheels to

Rogers Locomotive Works became America's locomotive giant, boasting five railroad engines built in a single year. By 1850, more than one hundred were made in a year. *Photo by Marcia Dente.*

This building originally was the Rogers Administration building. In 2003, it became the Frank R. Lautenberg Transportation Opportunity Center, which helps low-income families. *Courtesy of the Paterson Museum.*

ease movement on curves, thus again lessening the cost of railroads. Rogers's locomotives were sold in Canada, Australia and British territories.

In 1837, Rogers designed and built the Sandusky, which contained his own design innovations. The Sandusky was placed in service in Ohio. It was built for the Mad River & Lake Erie Railroad. As Rogers's reputation for producing locomotives of endurance and increasing power grew, more orders arrived, and the firm established itself in an important position in the industry. Perhaps the most well-known locomotive produced by Rogers was that bearing the serial number 31. Built in late 1855, the locomotive was purchased by the Western & Atlantic Railroad. Christened the General, this locomotive would become famous during the Civil War for an attempt by Union cavalry to highjack the Confederate train it was powering. The event was popularized in the 1962 movie *The Great Chase*, and the General is preserved today at the Southern Museum of Civil War and Locomotive History in Kennesaw, Georgia. A Rogers's locomotive (Union Pacific No. 119) that was built in 1868 pulled the special train that carried Union Pacific vice-president Thomas C. Durant to Promontory Summit for the golden spike ceremony that marked the completion of the first continental railroad on May 10, 1869. The original locomotive was a 440 American, but sadly, the original locomotive was scrapped for its iron in 1901. A replica

"Puffy" is the oldest continuously operating steam engine in the United States, built in 1896 in Paterson for Los Angeles, California. It is still used today. *Courtesy of Grapevine, Texas.*

was built in 1979, the first steam locomotive built in the United States since the late 1950s. The Grapevine Vintage Railroad's steam locomotive No. 2248, affectionately known as Puffy, was called a "heavy mountain-class locomotive," as indicated on the original Cooke drawings. Built in 1896 at the Cooke Locomotive Works in Paterson, it was originally purchased by the Southern Pacific Railroad. It was later converted into a fire train and then used as a ceremonial engine before ending up in a private collection, where Walt Disney began eyeing it for a railroad project that never came to be. Puffy helped launch the Texas State Railroad Excursion in 1976 and eventually the Fort Worth & Western Railroad's local Tarantula run. It made its first trek to Grapevine in 1993 and was used on special occasions to pull 1925-era passenger cars along the historic Cotton Belt Route between Grapevine and the Forth Worth stockyards.

The Passaic Rolling Mill, founded by Walter Cooke, built bridges for railroads by making a toy wooden model and then the full bridge, which was assembled without fastenings in the mill yard and was then dissembled, shipped and erected at the site.

By 1854, the Rogers firm (Rogers, Ketchum and Grosvenor) had made 103 engines, and by 1873, it was producing 1 engine every second day on an assembly line. Thomas Rogers died in 1856, and his son, Jacob, inherited the locomotive works. Jacob took the helm and reorganized the firm into the Rogers Locomotive and Machine Works. The company maintained its competitive position in the industry and prospered. After a huge fire in 1879, Jacob built the structure that is still standing today on the corner of Market and Spruce Streets. It was used as an erecting shop until Jacob suddenly closed the works and sold them to the American Locomotive Company in 1901. In the early 1890s, Jacob Rogers resigned the presidency but remained an investor, and the company was reorganized under its former treasurer, Robert S. Hughes, as the Rogers Locomotive Company. Hughes died in 1900, and the works were closed in 1901 by Rogers, who died later that year. Rogers left much of his fortune ($5 million) and collection of valuable artwork to the Metropolitan Museum of Art in New York City.

In 1832, Matthias W. Baldwin of Philadelphia produced his first locomotive, Old Ironsides, which was used on the Philadelphia, Germantown and Norristown Railroad and stayed in service for twenty years. His locomotive works ultimately became the largest in the United States, producing over 70,500 locomotives when it ceased operations in 1956.

William Swinburne was born in Brooklyn, New York, but settled in Paterson in 1833. Two years later, he was employed by the Rogers Locomotive

This is an interior view of the Rogers Locomotive and Erecting Shop taken between 1905 and 1915. It now houses the Paterson Museum. *Courtesy of the Library of Congress.*

Danforth, Cooke and Company was a competing company for the Rogers Locomotive Company. Eighty percent of all American steam locomotives were made in Paterson. *Courtesy of the Paterson Museum.*

Works as superintendent and patternmaker and assisted in the construction of Rogers's first locomotive, the Sandusky, in 1837. In 1848, he left to form his own locomotive works in partnership with Samuel Smith. What turned out to be the smallest of the three main locomotive works that began as a partnership between Smith, Collier and Bradley in 1842 eventually became Swinburne, Smith and Company. The company began to build locomotives and eventually began to enlarge. In 1851, the name was changed to the New Jersey Locomotive and Machine Company. Swinburne, Smith and Company was one of about twenty locomotive concerns that went under a decade later in the 1857 financial panic. It soon afterward was reorganized and purchased by the New York & Erie Railroad as a repair shop. After this time, Swinburne retired from the machine business and devoted his energies to the city of Paterson by holding several public offices. Another employee, John Cooke, who was born in Montreal, Canada, apprenticed at the Rogers Locomotive Works at the age of fifteen. He was made superintendent of the shop and held that position until 1852, when he joined Charles Danforth to form a new locomotive and building firm. Danforth, Cooke and Company incorporated in 1865 and was named after its principal partners. It soon became a serious competitor of the Rogers shop. Cooke was directly in charge of locomotive production, and Danforth was an expert in the area of textile machinery. In 1871, this firm changed its name to Cooke and Company and was purchased by the American Locomotive Company shortly after the turn of the century. It produced close to three thousand units before closing in 1926. During the

Grant Locomotive Works built approximately 1,888 locomotives for Paterson, as well as the city of Chicago, between the years 1867 and 1895. The company built locomotives for at least 134 railroads. *Courtesy of the Paterson Museum.*

Cooke Locomotive Works manufactured steam railroad locomotives starting in 1852 until it merged with other locomotive companies to form the American Locomotive Company in 1901. *Courtesy of the Paterson Museum.*

This photo is a detail of the Grant Locomotive Works Erecting Shop on Market and Spruce Streets. This was the smallest of the three locomotive works in Paterson. *Courtesy of the Library of Congress.*

late nineteenth century, Paterson was establishing itself as a major center for locomotive manufacturing in the country. Between 1863 and 1864, Oliver DeForest Grant bought stock in the company and, in 1867, changed the name to the Grant Locomotive Works.

Reorganized once more, the plant reopened for a brief time but could not compete with a newer conglomerate, the American Locomotive Company (ALCO), or its older rival and consistently leading U.S. manufacturer of locomotives, the Baldwin Locomotive Works of Philadelphia. It was finally absorbed into ALCO before the end of the decade, joining the Cooke Locomotive and Machine Works.

In 1901, the works were sold to the International Power Company, which sold out to the American Locomotive Company. Where there had been several locomotive shops, there was now one. ALCO continued making locomotives at the Rogers's plant for a few more years until major locomotive production and an important era in Paterson's history came to an end.

A Cooke 260 Mogul engine named No. 299 had been sitting on display at the Balboa Heights railroad station in the Panama Canal Zone since 1955. Looking for a centerpiece for the Historic District, Paterson's mayor, Lawrence "Pat" Kramer, negotiated the return of the engine to the city of

Engine No. 299, built in 1906, was used to build the Panama Canal and returned home to Paterson on June 6, 1979. It is the centerpiece of the Paterson Museum. *Photo by Marcia Dente.*

Paterson. He stood on the docks at dawn in the Canal Zone and watched as Engine No. 299 was placed on a giant tender for its final passage through the locks of the Panama Canal. At the end of the canal trip, it was hoisted onto a freighter and brought home to Paterson. It was given a citywide welcome with a parade, and it is now on display in the yard at the Paterson Museum.

The mills in Paterson can be seen in a variety of ways. They can be admired for their handsome architecture, or they can be viewed as monuments of America's Industrial Revolution. They are places where people spent a large portion of their lives. The mills were the heart of Paterson's industry. Without these buildings, locomotives could not have been built or silk cloth woven. All these structures were originally built for a specific purpose. The large doors on the first floor of the Thomas Rogers Building (now the home of the Paterson Museum) were built so that the giant locomotives could be moved out into the street after they were completed. The height of the ceiling is another indication of the size of the locomotives that were built there. The textile mills have their own character and structure identity. The Essex Mill was completely rebuilt in 1870 and represented the ultimate in modern plant design. It was the first fully integrated mill in Paterson. The raw cotton entered at one end, and the finished product exited at the other.

The brick and granite mills that now compose the Great Falls Historic District are handsome reminders of Paterson's past as an industrial leader.

JOHN PHILLIP HOLLAND

An unusual lozenge-shaped vessel was brought to the Passaic River's banks in July 1878. John Phillip Holland was about to launch his first submarine. Building a vessel that could submerge and move underwater had long been a dream of the Irishman.

John Phillip Holland was the inventor and designer of the first submarines in American, British, Japanese and Dutch navies. He was born on February 24, 1841, in Liscannor in County Clare and joined the Christian Brothers in 1858 before leaving the brotherhood in 1873. Brother Phillip, as he was known, taught at the North Monastery School between 1858 and 1861. While at North Monastery, Holland designed and prepared the drawings and specifications for the first submarines. Brother James Dominic Burke made a wooden model for him. The first tests were carried out with models and explosives in the monastery's ornamental pool.

John Holland was an Irish-born American submarine designer and founder of the American Holland Torpedo Boat Company. He also set up a company in Europe in 1899. *Courtesy of the Paterson Museum.*

Never an exceptional teacher of the classics, he found it almost "intolerably boring" to teach reading, writing and arithmetic, but he was known to be an excellent teacher of drawing, science and music. However, he never took full vows in the Order. While teaching in Enniscorthy, Holland developed an interest in flying and began to design flying machines. Between 1865 and 1869, he constructed a mechanical duck, which resembled an ordinary duck and could walk around, swim, dive and come to the surface again when put in the water. He was constantly devising mechanical contraptions. (There is still a sundial in the gardens of the monastery that Holland constructed and a pane of glass on which he etched the Sacred Heart in color.)

A combination of ill health and his inability to teach, as well as both his mother's and brother's immigration to the United States, resulted in Holland's declining to take his vows to the Order. On May 26, 1873, he, too, immigrated to the United States.

In New Jersey, Holland resumed teaching for a time at St. John's Parochial School in Paterson and then became involved with the Irish Fenian Brotherhood, which financed the building of his first three submarines. Holland, an Irish nationalist, and the Fenians, a nationalist group of Irish revolutionaries, were looking for a way to harass the British navy. John built a small prototype submarine called *Holland I* to test out his ideas, including the use of a gasoline engine. On May 22, 1878, the submarine was brought down to the Passaic River and launched from Lister's Boathouse above the Great Falls before a small audience. However, someone forgot to insert two screw plugs, and the submarine promptly sank to the bottom of the river just

Living so close to Westside Park, my parents always took pictures standing in front of the Fenian Ram before it was moved to the museum. This is my mother and me on New Year's Day 1951. *Photo by Harry Dente.*

above the Spruce Street Bridge. The hull of the vessel, minus its turret, was recovered fifty years later by a group of resourceful young boys using a magnet to locate the scuttled hull in the river's mud. A plaque on the bridge at the end of Front Street describes the exploits of these youths. They presented *Holland I* to the City of Paterson in 1927, and it is currently on display at the Paterson Museum. When the submarine was recovered, Dominic Conti, the owner of a sandwich shop and grocery store on the corner of Mill Street serving tubular-shaped sandwiches, thought it looked like the sandwiches he sold at his store, so he began calling them "submarine" sandwiches.

Holland's trial was successful enough to encourage building a larger, more warlike boat. In 1874, he submitted a submarine design to the secretary of the navy, who passed the paperwork on to a subordinate. Even if the idea of an underwater craft had any merit, the subordinate officer warned Holland that "[putting] anything through Washington was uphill work."

In 1881, Holland's submarine stayed underwater for an hour. Because it lacked a periscope, the submarine collided with the Weehawkin Ferry and sank. (The submarine had to broach the surface so the crew could look out through the windows in the conning tower.) Holland persisted and eventually incorporated the periscope once it was invented by a contemporary, Simon Lake, from Ocean City, New Jersey.

Holland launched the *Holland II* (nicknamed the "Fenian Ram") in May 1881. It was a thirty-one-foot, three-man boat, armed with a ram bow and

John Holland developed the modern submarine fitted with a petrol engine for surface propulsion and electric storage batteries and motor for running submerged. *Courtesy of the Paterson Museum.*

an air-power cannon. The vessel carried Holland and others to depths in excess of forty-five feet and successfully fired projectiles. Tests continued for two years, with the craft reaching depths of sixty feet for as long as one hour. Surface and submerged speeds were about the same, nine knots. The Fenians, however, became increasingly frustrated with Holland's delays and, faced with some internal squabbles, stole their own boat in 1883 but shortly realized that nobody but Holland knew how to work it. They hid it in a shed in New Haven, Connecticut, where it remained for thirty-five years. Holland had nothing more to do with the Fenians after that.

As a monument to Holland, Edward Browne purchased the Fenian Ram in 1927 and had it placed on display in Westside Park, in the Totowa section of Paterson. It remained there until 1980, when the city administration and the board of education decided to take the park property and build a new high school, John F. Kennedy High School. The submarine and Holland's papers were then moved indoors to the Paterson Museum.

The *Holland VI* eventually became the first submarine in the United States Navy. Purchased by the United States government on April 11, 1900, the USS *Holland* was commissioned on October 12, 1900, and served for ten years. It was considered the world's first practical submarine, and its acceptance by the navy marked the beginning of the United States submarine force.

In 1928, the Fenian Ram was put on display in Westside Park and then was moved to the Paterson Museum in 1980. *Courtesy of the Paterson Museum.*

By 1905, Holland had withdrawn from the Electric Boat Company that he had helped found and where he built his vessels, but not before the submarines were being used by American, British, Japanese, Dutch and Russian naval forces. The Electric Boat Company was established in 1899 to complete a fifty-four-foot submersible vessel developed by Holland. A submarine prototype that predates the Fenian Ram is the *Intelligent Whale*, stored in Sea Girt at the National Guard base. It had no engine, was hand cranked and is thought to be Holland's inspiration for submersibles.

Holland eventually moved to Baltimore, where he became very successful. He died of pneumonia on August 12, 1914, in Newark, New Jersey, and is buried in Holy Sepulchre Cemetery in Totowa, New Jersey. Regarded as the father of the modern submarine, Holland was remembered at North Monastery in County Cork, now celebrating two hundred years in existence. They dedicated its library at Ringskiddy, Ireland, in Holland's memory on May 7, 2012. The library has a unique collection of John P. Holland papers. Bruce Balistieri, curator of the Paterson Museum, gave the keynote address at the dedication.

Wright Aeronautical Corporation

The Wright Aeronautical Corporation moved to Paterson in 1921 and initially occupied space in a small, old silk mill. For a time, airplane engine production became the city's primary industry. It manufactured the single air-cooled Wright Whirlwind J-5C nine-cylinder radial engine that powered Charles Lindbergh's *Spirit of St. Louis* on its epic May 20, 1927 transatlantic flight. The plant underwent expansion after its "whirlwind" motor powered Lindbergh's plane. The motor is one of Paterson's major success stories.

"Our messenger of peace and goodwill has broken down another barrier of time and space," President Calvin Coolidge said about Charles Lindbergh's solo flight in 1927. No other aviation flight would generate as much excitement until the 1969 moon landing of Apollo 11.

In 1922, Charles Lindbergh left the University of Wisconsin after a year and a half to study aeronautics with the Nebraska Aircraft Corporation. He was a "barnstormer" until 1924, when he enrolled as a flying cadet in the Army Air Service. He won his reserve commission and began serving as a civilian airmail pilot, flying the route between St. Louis and Chicago.

In early 1927, he obtained the backing of several St. Louis men to compete for the $25,000 prize offered by New York hotel owner Raymond Orteig.

Wright Aeronautical Corporation moved into an old Paterson silk mill in 1921. It merged with Curtiss Aeroplane and Motor Company in 1929. *Courtesy of the Paterson Museum.*

Workers are inspecting the Whirlwind J-5C engines made on an assembly line. Used over many years, on thousands of airplanes, it only weighed 520 pounds. *Courtesy of the Paterson Museum.*

Several pilots were killed or injured while competing for the prize, and by 1927, it still hadn't been won. In February of that year, Lindbergh placed an order with Ryan Airlines in San Diego for an aircraft with specifications necessary to make the flight. The events leading up to the flight involved months of planning. Lindbergh's concern with every detail helped ensure his success. The aircraft was completed in April 1927 and was powered by a reliable Paterson-made 223-horsepower Wright Whirlwind J-5C engine. Lindbergh made several test flights before flying the aircraft from San Diego to New York on May 10–12, making only one stop at St. Louis. His flight time was twenty-one hours and forty minutes. It not only won Lindbergh the prize, but it also set a new transcontinental record, was an aviation breakthrough and won Lindbergh worldwide celebrity.

Equipped with four sandwiches, two canteens of water and 451 gallons of gas, Lindbergh took off for Paris alone on the morning of May 20, 1927. He landed safely at Le Bourget Field near Paris thirty-three hours later, where he was greeted by a wildly enthusiastic crowd. Halfway through the flight, "Sleet began to cling to the plane. That worried me a great deal and I debated whether I should keep on or go back. I decided I must not think any more about going back," Lindbergh said.

Right: In 1927, Charles Lindbergh and Donald Hall designed the *Spirit of St. Louis* and planned the flight to Paris in sixty days. *Courtesy of the Paterson Museum.*

Below: The truck that transported the Wright J-5C Whirlwind engine that powered Lindbergh's plane. It was estimated to be able to perform flawlessly for over nine thousand hours. *Courtesy of the Paterson Museum.*

On April 30, 1928, after taking his famous plane on a tour of the United States, the *Spirit of St. Louis* made its final flight, traveling from St. Louis to Washington, D.C., where Lindbergh presented the aircraft to the Smithsonian Institution.

On May 21, 1927, Paterson joined in the celebration of the solo flight by carrying a similar engine on the back of a pickup truck through the streets of Paterson and sitting it in front of the Hamilton Club on display. In 1929, the Wright Aeronautical Corporation merged with Curtiss, and the Curtiss-Wright Corporation was formed. By 1938, it was the largest airplane engine factory in the United States, with more than 3,500 employees producing over 2,000 engines. The company experienced enormous growth and achievement during World War II, which facilitated its exodus from its Paterson operation in 1945. In fact, many Paterson factories contributed to the production of aircraft engines for many World War II bombers (engines for the first B-17 Flying Fortresses) and for the *Enola Gay*, which dropped the atomic bomb on Hiroshima and effectively ended the war in the Pacific. Jimmy Doolittle raided Tokyo in daylight in planes powered by engines made in Paterson, and the city's factories produced nearly 1,040 aircraft engines, surpassing all other American cities in production.

On the fifty-eighth anniversary of Lindbergh's flight, a plaque was dedicated at the factory on Lindbergh Place (formerly Lewis Street) in a ceremony with Paterson mayor Frank X. Graves Jr. and then-congressman Robert A. Roe while a vintage aircraft flew overhead.

Chapter Seven

Silk City of the World and the Silk Strike of 1913

Silk, long known as the queen of textiles, is valued for its luster, soft feel and ability to be hung up. It is used for both clothing and home décor, as well as ribbons and braids. Silk filaments are harvested from the silkworm, which feeds on the leaves of mulberry trees. While world production of silk goes back more than 4,500 years ago to China's first production, the silk industry in what is now the United States began in 1603. Since America's climate was better suited to silk culture than that of Britain, King James sent silkworm eggs and mulberry tree seeds to the new colonists of Virginia. Cotton and tobacco proved to be more profitable, and silk production was somewhat limited from 1603 to 1760.

As early as 1794, SUM considered the propagation of white mulberry trees as a potential gateway into the wider silk industry but did not act on this idea. Few early American silk mills were established before the 1860s. These early efforts were curtailed by the difficulty in obtaining suitable raw silk and affordable skilled labor.

During the 1830s, something like a fever gripped the country as the cultivation of mulberry trees and the husbandry of silkworms was promoted as a fast track to riches. What was overlooked was the care and feeding of the silkworm, which was difficult work. In the early 1840s, after the craze had waned, a mulberry tree blight drove most of the remaining silkworm dreamers from the business.

Silk was manufactured in Paterson in 1840 but did not prosper until after the Civil War, when high tariffs on imported silk products helped American producers compete with their European rivals. Silk manufacturing was a big business before the days of synthetics. Silk dominated high fashion and the luxury trade, but women at all economic levels wanted their best dresses to be made of silk. Paterson had many advantages for the silk trade. It had abundant water supplies for power and processing and good transportation facilities. It was close to New York City, the center of the fashion industry. Most importantly, Paterson had a supply of workers who understood the peculiar characteristics of the delicate silk fiber. By the 1880s, the city was producing almost half of the silk manufactured in the United States and had become known nationwide as "Silk City."

For more than a century, American silk manufacture has been steadily developing itself. Various circumstances have helped or hindered, but none has interrupted its growth. A superficial glance at the history of the silk industry in this country reveals three figures who were its leaders and in whose footsteps hundreds of others have followed. John Ryle, Catholina Lambert and William Strange will always be remembered for bringing the silk industry to its present flourishing condition.

Of all the industries in Paterson, silk has always been predominant. It didn't matter if the nation was prospering or if people were falling on hard times, there was always a demand for silk, and Paterson could always depend on silk. There were thirty-two silk-manufacturing companies and five silk-dyeing shops in Paterson by 1875, and by 1900, over twenty thousand workers were employed in Paterson's silk industry. By 1913, one-fifth of Paterson's residents were employed in the silk industry. Not all workers were content, however. Hours in the factories were long, conditions were challenging and the pay for unskilled work was low. Many workers were paid only "by the piece," and they needed to work extraordinarily long hours in the factory as well as at home in order to make a decent week's wages. Additionally, child labor was rampant. Factory owners needing to compete in an increasingly mechanized industry tried to squeeze as much time as possible out of their workers, and their efforts to survive in an increasingly competitive market included strategies that required workers to be more productive.

For many years, the Falls was made to turn the wheels of industry. The area served to attract visitors with its picturesque beauty, and thanks to the energy and public spirit of silk manufacturer John Ryle, that natural beauty has been measurably preserved in a park.

Although Christopher Colt, Sam Colt's father, was the first to introduce silk manufacturing to Paterson (though he produced little or no silk in the Gun Mill), he is not credited as having been the founder of nineteenth-century Paterson's largest industry. That honor is bestowed on John Ryle.

Ryle was born in England at the epicenter of the British silk industry. At age five, he worked as a "bobbin boy" in a silk mill, and at age twenty-two, he was superintendent of a mill owned by his brother. In 1839, he left England for America and first took employment in a silk mill in Massachusetts. Late in 1839, George W. Murray persuaded him to come to Paterson and evaluate the machinery Colt left behind in the Gun Mill. Murray had paid Colt $3,200 for his machinery and hired Ryle to oversee putting the enterprise into effect. In 1843, Ryle's initial contract with Murray ran out. At this point, he purchased a share in the business, which then became known as Murray and Ryle. In 1846, the partnership occupied both the third and fourth floors of the Gun Mill. Ryle paid $600 a year in rent for these two floors (the first and second floors remained vacant due to the failure of the Patent Firearms Manufacturing Company). That same year, Ryle was able to buy out all of Murray's interest in the business. Assuming full control, Ryle installed skylights in the fifth floor and began weaving fine broad silk, another first for Paterson. The most famous product of Ryle's looms was the twenty- by forty-foot American flag that flew above the Crystal Palace in New York City.

It was Ryle who was the first to put silk on a spool. In response to a plea from his friend Elias Howe, he found a way to put silk onto spools for Howe's sewing machines. This enabled Howe to overcome one of the chief difficulties he had in perfecting his sewing machine; the silk thread could now be fed to the needle. Ryle's machine twist was the first of its kind that could be used on a sewing machine successfully. This was the beginning of the spool silk industry in this country and the beginning of Paterson's renown as the "Silk City."

By 1854, Ryle's complex of buildings had expanded to include a row of stone buildings that contained additional silk-manufacturing facilities and a dye house. In the early days of Paterson's silk industry, product had to be sent to Philadelphia to be dyed because there was not a dye house. Ryle built a dye shop in the Gun Mill yard to solve this problem. In 1858, the first custom dye house in Paterson was established by John Mayer and William Brown. Independent from Ryle, they occupied a building in the Gun Mill yard. For as long as it retained its full five-story height and its graceful bell tower, the Gun Mill remained one of the most memorable sights in Paterson. The silhouette of the tower rising above the river gorge of the Great Falls was unusually striking.

Ryle's interests and responsibilities did not end with the manufacture of silk. He was, for a time, the largest shareholder and the driving force behind the Paterson Water Company. He rose to become a well-respected civic figure and served as mayor of his adopted city from 1869 to 1870.

Although a handful of other attempts at silk manufacturing were made in Paterson in the 1840s, there were no other large companies that compared with Ryle's silk works. The high cost of American labor and difficulty acquiring raw silk of any quality stunted the growth of the industry. Most of the mills that opened subsequently failed.

Ryle reorganized over the years. His uninsured Murray Mill was consumed by fire and reorganized under the guise of the Ryle Silk Manufacturing Company. It, too, became insolvent and, a few years later, was replaced by John Ryle and Sons, which quickly merged into the Pioneer Silk Company that survived until 1901.

In 1860, Paterson was home to six silk companies, employing six hundred workers. One-third of the city's total workforce (about seventy-three thousand people) was employed in the silk business. The lure of work and the desire to escape poverty brought immigrants, including skilled workers, from England, France, Italy, Poland, Germany and Russia.

The last years of the Civil War saw the establishment of the Barbour Flax Company here. The parent company in Ulster, Ireland, had practically a world monopoly in the production of linen thread. The new American complex of five mills soon led the United States in the output of that product. It was the largest linen works in the country and, in the second half of the nineteenth century, employed 440 people to process over 1.5 million pounds of flax annually.

At the turn of the century, a radical turning point was sharply felt in Paterson. Jacob Rogers closed the Rogers Locomotive Company, signaling the beginning of an industrial retreat that became a route. If railroading was dying, silk was not. Mill owner Catholina Lambert's nouveau riche castle, Bella Vista (named after his wife, Isabelle, and for the spectacular view of the Manhattan skyline from its terrace), now provides a surprisingly Gothic home for the Passaic County Historical Society. Lambert himself had been a poor English immigrant who had worked a seventy-two-hour week at the age of ten in a cotton mill in England. In 1851, at the age of seventeen, he left England and found work as a bookkeeper in a Boston textile firm, Tilt & Dexter. Four years later, he became a partner and head of the company three years after that, when he then relocated many of its interests to Paterson. Dexter, Lambert & Co. grew into one of Paterson's most important silk manufacturers during the 1860s and 1870s. By 1890, he had become one of the largest mill owners in Paterson. He made enough money from his five mills to pay for his English-style castle and his collection of some four hundred paintings. It is estimated that the castle cost $500,000

Lambert Castle, originally called Bella Vista and built by Catholina Lambert, is one of the state's last estates that date to America's Gilded Age. *Courtesy of the Paterson Museum.*

to build in 1892, at a time when the average wage was $1 a day for ten hours or more of hard labor. Lambert lived in splendor while his weavers endured miserable working conditions.

The first strike in Paterson was in 1794 and was the result of the absence of adequate schools for workers' children. The first factory labor strike in the new nation occurred in Paterson on July 19, 1828, when women and child millworkers quit their looms, demanded a twelve-hour workday and protested a change in the lunch hour. The mill owners had decided that it would be better for the child workers to take their midday meal at one o'clock rather than at noon, feeling there would be a more equal division in the workday. Employees surprised management by demanding the reduction of their workday hours from thirteen and a half to twelve. In support, other local workers in the city such as mill hands, journeymen artisans, carpenters, masons and mechanics walked out. This was considered the first sympathy strike recorded in the United States. The workers finally lost the strike, but it made a strong impression on the community. The owners eventually restored the noon lunch hour. It was nicknamed "the children's strike" and lasted three weeks.

In the 1830s, Paterson's factory workers attempted to change and improve working conditions, especially for children. The strike of 1835 involved over two thousand workers and lasted six weeks. At issue were the twelve- to fourteen-hour workdays for children, with only a half-hour breakfast and a forty-five-minute lunch break. Parents demanded reducing the children's workday to eleven hours. Paterson workers received support from workers' organizations in New York, Boston and Newark, and workers turned to unions for the first time. The strike succeeded.

The Panic of 1873 and the resulting depression crippled Paterson's economy. Strikes and protests preventing wage cuts were commonplace. In the 1877 summer months, Paterson was hit by an unprecedented strike involving two thousand ribbon weavers. The weavers successfully reached an agreement to get a pay cut reversal. The mill owners learned a powerful lesson: the community, including elected officials and the police department, was sympathetic to the strikers. Mill owners would spend the ensuing years gaining influence in municipal government and establishing their own private police forces.

During the late 1800s, a three-hour strike led by foreign workers took place in local textile factories. The strike was held to protest unbearable conditions in the silk mills but did little to change the miserable working conditions. Somehow, organized labor had never been successful in Paterson. From 1879 on, there were a number of textile unions, most of them independent.

A typical day for unskilled workers meant twelve hours standing at a machine doing boring, repetitive work, with only a twenty-minute break for lunch. Conditions at various mills were described as "filthy, dingy, damp and dank, and just short of inhumane." Workers were paid little more than one dollar a day during their first few years at the mills. Sexual harassment laws did not exist, and some bosses were of a sleazy character; others were just downright mean.

Between 1901 and 1913, Progressive legislators courting the labor vote gave New Jersey workers increased control over wages, hours and working conditions. Paterson's textile unions added thousands of workers to their organizations. Union laws were able to negotiate wage and price agreements with the large mills and improve working conditions in the smaller shops. Child labor laws were passed limiting children to eight-hour workdays and eliminating work on Sundays.

The panic of 1912 brought bankruptcies, pay reductions and mill closings. In 1913, Paterson was the site of the greatest textile strike in American history, one that captured national attention, and it eventually expanded to include fifty thousand workers in New Jersey, Pennsylvania, New York and Connecticut. It was a bad year everywhere. Strikes were rampant in the mines and railroads and also among construction workers. The most radical union, the IWW (International Workers of the World), had scored triumphs in the Oregon lumber camps. Suddenly, Paterson became the vortex of it all.

Years of smoldering resentment over low wages and miserable working conditions culminated in the weavers' outrage at new looms that would cut jobs in half by reducing the workweek from fifty-five to forty-four hours. The invention of rayon in 1912 also doomed the silk industry.

It started in the Henry Doherty Silk Mill on January 27, 1913, when on a cold morning, eight hundred broad silk loom weavers walked off the job in protest of the firing of their negotiating committee. The strike soon involved practically the whole silk industry, some twenty-five thousand workers. The entire industry was at a standstill.

Henry Doherty immigrated to America in the late 1860s and worked as a foreman for ten years in Paterson silk mills and then began a one-loom operation in a rented space. By the early 1890s, his firm employed nearly one thousand workers. The Doherty Plant was one of Paterson's largest and most modern silk mills. When the weavers walked out of his mill, Doherty had good reason to think that they would soon be back, as they had been in the past. Strikes disrupted production and reduced profits, but they were all a part of "being in silk." The workers' grievance centered on Doherty's extension of the four-loom system, and the new looms were equipped with automatic controls that stopped the machinery whenever a thread broke. One weaver could now tend to three or four looms rather than two, and every other weaver would be thrown out of work. But job security wasn't the only issue. Competition from low-wage silk mills in Pennsylvania and the South prompted Doherty to turn to the four-loom system to increase productivity and save jobs. He refused to discontinue the system he had invested a great deal of money in. Catholina Lambert was adamantly opposed to dealing with the strikers. When Henry Doherty was prepared to agree to a settlement with the workers, Lambert angrily confronted him and reportedly had to be restrained from physically attacking him. However, Lambert could not recover from the strike and was forced to declare bankruptcy.

The workers were also upset over long hours and poor working conditions. The protest of the weavers provided a spark. Soon all the men, women and children who worked in the mills also rebelled, so the workers took their discontent to the streets. The silk manufacturer was justified in thinking that this strike, like the others, would peter out as soon as the workers exhausted their meager savings. He had good reason to believe that, but Doherty was wrong. It is unclear how the Paterson strike was transformed from a walkout at a single mill into a general strike that shut down more than three hundred mills and dye houses, but the Paterson Silk Strike of 1913 was underway. The general strike was a new development in the industry. For the first time, the mill owners faced a united front. They could not fire the entire workforce and depend on a steady stream of new immigrants eager to work at any wage to fill the unskilled jobs. This time the strikers were the new immigrants. Also in 1913, the leadership was different. In past strikes, the mill owner had negotiated with longtime employees he knew well and with whom he concluded gentlemen's agreements to settle the dispute. Unions were not recognized. In this strike, the mill owners were confronted by the IWW, the most radical union in the United States.

The labor troubles that descended on Paterson in the winter of 1913 were nothing new for the residents of the "Lyons of America." Between 1880 and 1910, the city recorded at least 140 strikes and work stoppages. Striking was the laborers' chief weapon against their capitalist employers. Some of the strikes became famous. In 1892, steel workers struck Andrew Carnegie's giant Homestead Mill in Pittsburgh, Pennsylvania. In 1893, Eugene V. Debs and railroad workers boycotted railroad cars built by the Pullman Company. A vast majority of the strikes during this period have been forgotten, but it was a time of widespread worker unrest, and at one time or another, most workers walked a picket line.

The broad silk weavers were soon joined by ribbon weavers and dye house workers. Nearly three thousand arrests were made, some for nothing worse than trying to hold meetings. Committees were established to handle communications, to manage a strike relief fund, to provide legal help for the arrested strikers and to organize mass meetings. Violence broke out among strikers, police and strikebreakers. Paterson workers bonded with the IWW labor union, also sometimes called the Wobblies. This union was small in size but infamous due to the socialist leanings of its primary leaders. Grievances were exacerbated by the awesome presence of the IWW's most famous national leaders, men like William "Big Bill" Haywood, Carlo Tresca and others. Emma Goldman, Eugene V. Debs and Elizabeth Gurley Flynn arrived in Paterson in mid-February, and all three came in defense of Paterson's defenseless workers. Flynn was put in charge of the female strikers. She publicized women's working conditions, which included poor sanitation, lack of heating, lack of compensation for serious injuries and sexual harassment. While some skilled women earned good wages, many others were subjected to fines that reduced take-home pay. At one mill, girls were required to sign a year's contract that allowed the owner to hold 50 percent of their pay until the end of the year, without interest. This amount was forfeited if they left his employment before the year was out. These three labor leaders helped organize a central strike committee to unite the various craft and ethnic groups, but they were of no avail. On February 23, 1913, the executive committee issued the call for a general strike, strategically calling it in order to interfere with the production of the spring fashion season. Nearly three hundred mills and dye houses were closed, and twenty-four thousand men, women and children joined in an industry-wide strike.

The strike immediately divided the Paterson community. Mill owners, managers, the mayor, chief of police, clergy and most of Paterson's middle class were on one side, and workers led by their own people and the IWW militants were on the other. This division was a new development in Paterson's industrial relations. In previous strikes, workers had enjoyed support throughout the community. This change was a major factor in the extraordinary length of the strike. As of March 1, 1913, the battle lines were drawn.

The Paterson silk strike lasted from the end of February until around the middle of July. In the beginning, the workers formulated a list of demands that added to the original grievance against the four-loom system. They demanded an eight-hour day for all mill works; the abolishment of the three- and four-loom system in broad silk and the two-loom system in ribbon; restoration of the 1894 price list on ribbon (piecework wages in 1894 were higher than the 1913 price list); a 25 percent wage increase for all winders, quilters, pickers and other industry workers; and a minimum wage of twelve dollars per week for dyer helpers. Compared to today's standards, an eight-hour day or wage increases seem like mild labor demands.

To achieve these demands, the workers employed a host of tactics. They aimed to strike all the mills and wait for the owners of the smaller mills to face bankruptcy. They would then settle and, in the process, force the larger mills to follow suit. A second tactic was the traditional picket line. The purpose of the picket line was to prevent workers from entering the mills and to keep out the "scabs." The workers manned the lines throughout the duration of the strike. The Paterson Strike was remarkably free of violence against the scabs, as the IWW constantly admonished the workers to shun violence and told them time and time again that they killed the mills by stopping production and breaking off the profits. The IWW also attempted to spread the strike beyond Paterson to other silk-manufacturing plants. It's unknown how successful these efforts were. There were sympathy strikes outside Paterson, but they were short-lived and had no influence on developments in Paterson.

A major struggle in the history of America occurred when 2,500 striking silk workers shut down three hundred silk mills and dye houses in Paterson. *Courtesy of the American Labor Museum/Botto House National Landmark.*

A gathering point for the strikers and their leaders was the Botto House in Haledon, just over the city line from Paterson, now the American Labor Museum. Pietro Botto and his wife, Maria, were northern Italian immigrants who had worked in the mills (Pietro as a skilled weaver) for fifteen years until they were able to fulfill their goal of owning their own home and garden. Maria kept house and cooked for weekend guests. They opened their home to other immigrants from their district in Italy, who gathered at their house for socializing and recreation on weekends as they had since 1908, when the house was built. When the workers were forbidden to gather in Paterson, Pietro Botto permitted the use of his house as a meeting site for massive groups of strikers. Every Sunday from the beginning of March until July 20, Botto's house became the focal point for the strikers' mass meetings. The Sunday meetings were among the most important events of the strike. The rallies were held in Haledon because the workers could gather there without fearing police interference, harassment or arrest. Novelist Upton Sinclair was one of the well-known Socialist figures who spoke from the Bottos' second-floor balcony to the crowd in the streets below, which numbered twenty thousand people. The mayor of Haledon, socialist

Pietro Botto, his wife, Maria, and their four daughters pose for a family portrait. Maria cooked for over one hundred visitors on a typical Sunday. *Courtesy of the American Labor Museum/Botto House National Landmark.*

William Brueckmann, also spoke. Paterson authorities moved quickly to crush the strike by forbidding the strikers to hold public meetings. Alone among elected officials, Brueckmann defied the Paterson authorities and their police departments and upheld the constitutional rights of the strikers by giving them a safe haven in his town. His action marked the struggle that brought thousands of workers to Haledon and forced the city of Paterson to its knees. Haledon was most likely the place where the Paterson Strike Pageant was conceived and organized. The Bottos were blackballed from the silk industry for allowing their home to be used as a strike headquarters.

Today, the Pietro Botto House is a landmark commemorating a chapter in the long and hard struggle of the American worker, especially the immigrant worker, to participate in the material fruits of the Industrial Revolution. The American Labor Museum is housed in the 1908 Botto House. The house was placed on the National Register of Historic Places in Passaic County on July 30, 1974, and was declared a National Historic Landmark by the Department of the Interior in 1982 because it served as a haven for free speech and assembly for the striking laborers. In 1983, the house opened as a museum with restored period rooms, a labor and immigrant library, Old World gardens and changing exhibits.

The Botto House played an important part in the silk strike. It's where leaders of the IWW Union spoke to the striking workers from the house's balcony. *Courtesy of the American Labor Museum/Botto House National Landmark.*

Another tactic was holding constant meetings, which were held in Paterson's halls during the week, especially the Helvitia and Turn Halls. The meetings held in Haledon served several purposes. First, they presented the workers and the families with the opportunity to relieve the tensions and worries of the strike. The meetings had the atmosphere of a picnic. Men and women talked, and children played. Second, the Haledon meetings boosted morale. In Haledon, the workers were one voice and one will.

The strikers called for radical times, and to achieve their demands, the workers relied on maintaining industry-wide unity. They agreed that no one would return to work until all of the owners accepted their terms. Picket lines kept strikebreakers out of the mills and intimidated those who did go back to work.

The mill owners refused to talk to the union. They tried to pit old and new immigrants against one another and worked with police, while the courts forbade speeches and rallies by the strikers. The mill owners formed a manufacturers' association whose purpose was clear, and they stood united. The owners agreed that the best way to break the strike was to "hold out until the harsh reality of empty stomachs and crying children forced the workers to answer the mill whistle." They farmed out what work they could to their Pennsylvania mills and waited.

By the end of May 1913, the workers' solidarity had begun to crack. They instructed their representatives to vote to accept settlements with individual mill owners. At the same time, workers started to cross the picket lines, despite the pleas of the IWW. Without question, the most famous event associated with the strike occurred on June 7, 1913, when the IWW launched the Paterson Strike Pageant, which took place in Madison Square Garden. It was a powerful and innovative use of the pageant form. John Reed conceived the pageant to rally the strikers, provide vitally needed money for the relief fund and forge a link between the radical "new" intellectuals in New York and the working-class revolutionaries of the IWW.

On the morning of June 7, approximately one thousand Paterson men, women and children gathered at Turn Hall. The men wore suits, and the women wore black skirts and white blouses. John Reed explained the scenes and rehearsed the participants. At about 11:00 a.m., the workers paraded to the railroad station and boarded special trains for the city. When they arrived, they formed ranks and marched to Washington Square for a short rally. The procession sang songs ("La Marseillaise," "Internationale" and Paterson strike songs) and proceeded up Fifth Avenue to Madison Square Garden. At the Garden, Reed rehearsed his one thousand actors once again. After a lunch of sandwiches provided by workers from the garment district, the silk millworkers rested and waited for the curtain to rise. Although scheduled to begin at 8:30 p.m., the fifteen thousand attendees were not seated until after 9:00 p.m. because of the crowd's size, the largest in the history of Madison Square Garden.

On a stage more than two hundred feet long (longer than a city block; it is the largest stage ever built in New York) and capable of holding more than one thousand performers (on a stage bare of props), workers from Paterson's silk factories re-created the central events of their strike. This stage performance grew out of the massive strike that began in February and was pulled together by the IWW organizers in an attempt to raise funds.

Mabel Dodge later wrote, "For a few electric moments there was a terrible unity between all these people. They were one: the workers had come to show their comrades what was happening across the river and the workers had come to see it. I have never felt such a pulsing vibration in any gathering before or since."

Organizing the visual and performance aspects of the pageant took place in and around Greenwich Village, where most of the organizers lived and worked. The organizers structured the pageant episodes narrating the events of the strike: first the walkout, with workers singing "La Marseillaise," followed by violent clashes with police, then the shooting of an innocent

IWW leaders Pat Quinlin, Carlo Tresca, Elizabeth Gurley Flynn, Adolph Lenning and Big Bill Haywood pause for a portrait during the Paterson Silk Strike in 1913. *Courtesy of the American Labor Museum/Botto Hose National Landmark.*

bystander and the IWW's mass meeting and, finally, ending with the workers and the audience triumphantly singing the "Internationale," the anthem of international socialism. Lesser geniuses would have hired a hall and exhibited moving pictures of the Paterson strike.

The Paterson Strike Pageant was clearly an artistic success. Even conservative news journals hostile to the pageant itself were impressed by its simplicity and power. It was supposed to be nothing but a vivid record of history. While the workers were clubbed and shot by detectives and policemen, the mills remained dead. While organizers were persecuted, the strike continued, and still the mills were dead. While the pulpit thundered denunciation and the press screamed lies, the mills remained dead. No violence could make the mills alive—no legal process could resurrect them from the dead. Bayonets and clubs, injunctions and court orders were equally futile. Only the return of the workers to the mills could give the dead things life.

The following is the program that was handed out the evening of the pageant:

Program of the Paterson Strike Pageant

Scene: Paterson, N.J.
Time: AD 1913

The Pageant represents a battle between the working class and the capitalist class conducted by the Industrial Workers of the World (IWW), making use of the General

Strike as the chief weapon. It is a conflict between two social forces—the force of labor and the force of capital.

Episode One: Part 1. "The Mills Alive—the Workers Dead"

Episode One: Part 2. "The Workers Begin to Think"

Six o'clock on a February morning. The mill windows all aglow. The mill whistle sounds the signal to begin work. Men, women, old and young, come to work in the bitter cold of the dawn. The sound of looms. The beginning of the great silk strike. The striking workers sing "La Marseillaise," the entire audience invited to join in the song of revolt.

Episode Two: "The Mills Dead—The Workers Alive"

Mass picketing. Every worker alert. The police interfere with peaceful picketing and treat the strikers with great brutality. The workers are provoked to anger. Fights between police and strikers ensue. Many strikers are clubbed and arrested. Shots are fired by detectives hired by the manufacturers, and Valentino Modestino, who was not a striker or a silk mill worker, is hit by a bullet and killed as he stands on the porch of his house with one of his children in his arms.

Episode Three: "The Funeral of Modestino"

A scene with all the somber realism, the actors brought their thousands of sympathizers in the audience to such a high pitch of emotion that it was punctuated with moans and groans and sobs. A coffin containing the body of Modestino was borne across the stage, followed by the strikers in a funeral procession to the heavy tones of the "Dead March." As they pass, the mourners drop red carnations and ribbons upon the coffin until it is buried "beneath the crimson symbol of the workers' blood" and then listen to Carlo Tresca and "Big Bill" Haywood, who spoke as they had at the actual funeral in Paterson. [Tresca made an impassioned speech blaming murder on profit-hungry capitalists, for which he was later indicted.]

Episode Four: "Mass Meeting at Haledon"

Great mass meeting of 20,000 strikers. IWW organizers speak. Songs by the strike composers are sung by the strikers. They also sing the "Internationale," "La Marseillaise" and the "Red Flag." The audience is invited to join in.

Episode Five: Part 1. "May Day"

The May Day Parade. The workers of Paterson, with bands playing, flags flying and women and children dressed in red, celebrate the international revolutionary labor day.

Episode Five: Part 2. "Sending Away the Children"

The scene dramatizes the emotional moments when the women on strike hand over their children to temporary "strike mothers" in other cities that their parents might go on and fight and starve and struggle unhampered by their little ones. With all the details of farewell embraces and tears and, finally, shouts of enthusiasm breaking through the sadness of parting while Elizabeth Gurley Flynn makes a consoling speech to the weeping mothers and rouses their spirits once more to the blind determination to fight on.

Originators of the Paterson Silk Pageant hoped to reach out to the hearts and pocketbooks of the workers in New York. *Courtesy of the American Labor Museum/Botto House National Landmark.*

Episode Six: "Strike Meeting in Turner Hall"

The strikers, men and women legislate for themselves. They pass a law for the eight-hour day. No court can declare the law thus made unconstitutional.

On June 8, hundreds of thousands of New Yorkers read about the key events of the strike in detail in their newspapers. Newspapers had been reporting that the Paterson strike was broken, but now the pageant had shown the people of New York the truth. Later in the month, outside contributions to the strikers' relief fund began to grow. So in terms of its original purpose of publicizing the strike, the pageant was an overwhelming success.

Neither the pageant nor the strike ended on a triumphant note—both were doomed from the beginning. There had only been enough money to rent the Garden for a single night. The pageant lost money, and the strike ended in defeat after five months. Nonetheless, the pageant represented an important moment in the alliance between modern art and labor radicalism.

Mabel Dodge wrote that no one realized the fun of having placed the letters IWW ten feet high on each of the four sides of the Madison Square Garden Tower in bright red electric lights so they could be seen from one end of town to the other. Inside the Garden, Sloan painted the ninety-foot backdrop of the Paterson silk mill, "the windows aglow with artificial light in which the workers began their daily tasks."

In recounting the events of the pageant, Dodge acknowledged that everyone worked but her. Dodge's job as she saw it was to inspire and to raise money. As a wealthy Fifth Avenue heiress, she spent a lot of energy trying to convince the anarchists in her circle that she was a good capitalist.

However, as Bertram D. Wolfe pointed out, "It's hard to fill Madison Square Garden. The dollar- and two-dollar seats remained almost empty until workers and strikers were let in for free or at ten cents a seat. Instead of making money the pageant ended with a deficit." Unfortunately for all concerned, the Paterson Strike Pageant failed to raise enough funds through admission fees to make a significant contribution to the depleting strike fund. Furthermore, the event led to a bitter argument between the labor movement activists and the Greenwich Village arts radicals, with labor leaders accusing the villagers of taking the profits of the pageant. The accusations proved unfounded as the disappointing profit of $250 after expenses caused a great deal of ill will. Because the strike fund was unable to raise enough money and the workers were starved into submission, the strike was broken in August 1913. The five-month strike in Paterson's silk industry, however, is remembered as a showdown between the ideals of capitalism and socialism. Both new and old immigrants were represented in both sides of the strike, more often than not with new immigrant workers being labeled as the radicals and the immigrant industrialists being labeled as thieves.

Elizabeth Gurley Flynn defended the way the IWW ran the campaign. "What is a labor victory? I maintain that it is a twofold thing. Workers must gain economic advantage, but they must also gain revolutionary spirit in order to achieve a complete victory. For workers to gain a few cents more a day, a few minutes less a day and go back to work with the same psychology, the same attitude toward society is to achieve a temporary gain and not a lasting victory. For workers to go back with a class conscious spirit, with an organized and determined attitude toward society means that even if they have made no economic gain they have the possibility of gaining in the future."

In July, the central strike committee voted to endorse shop-by-shop settlements. The Paterson silk strike was over. The owners had conceded nothing. The workers lost $5 million in unpaid wages, and the mill owners lost $10 million in profits. After twenty weeks, workers were destitute and many mills were bankrupt. They refused to bargain, and not one mill was established after 1913. There is still debate about the effects of the strike. The owners agreed to none of the strikers' demands, and the two-loom system remained still the standard in Paterson in 1919. Paterson's large-scale silk manufacturing accelerated its transfer to Pennsylvania, and the defeat in Paterson marked the end of the IWW's effectiveness in the East.

Still, most all contemporary accounts indicate that, at least for the night of the Paterson Strike Pageant at Madison Square Garden, everyone was uplifted and in good spirits.

"The workers didn't go back to their jobs as the same people," notes Evelyn Hershey, education director of the American Labor Museum and Botto House National Landmark. "Many had a new found respect for workers in other crafts and from different ethnic backgrounds and many, especially the women, inspired leaders in many union battles to come." Perhaps most importantly, the Paterson silk workers showed that workers can overcome ethnic, racial and gender differences to achieve solidarity, and they set the stage for future union victories in other industries.

When Franklin D. Roosevelt became president in 1932, Eleanor Roosevelt chose silk from Paterson to wear to the Inaugural Ball. The first lady's gown, made of a metallic fabric from Rosenstein Brothers on Broadway, is in the Smithsonian Institution in Washington, D.C. But the great days of the Silk City and its famous product were passing. Beginning in the 1920s, rayon ("artificial silk") gained popularity and began to replace silk.

The industry had slowly been losing ground for years. Then on October 21, 1929, came the stock market crash, bringing a decisive end to the high-flying 1920s. As bank failures, unemployment, foreclosures and evictions spread across the country, the weight of the Great Depression fell on Paterson and the silk industry. In downtown Paterson, people from the dye houses stood in line at the soup kitchen. Since silk manufacturing had employed a large percentage of Paterson's wage earners in 1925, the decline of the industry, added to the effects of the Depression, left the city in deep distress. Between 1929 and 1933, union membership in the Paterson textile industry had been dwindling. At the depth of the Depression, it was negligible. The slow leak of silk mills out of Paterson continued. Firms that had plants outside the city were closing the ones in Paterson. The silk business was leaving Paterson. The whole industry was in decline.

At one time, Paterson had been considered a perfect city for textiles, good sewage, good water and good proximity to the market in New York City. But Paterson was no longer a good choice for textile manufacture. Its old multistory buildings adapted poorly to large-scale operations, and textile firms kept closing and moving to easier situations in the South. Paterson's title as Silk City had become a historical artifact.

The source for the pageant program in this chapter is from Frederick Boyd's *The Pageant of the Paterson Strike* (New York, 1913).

Chapter Eight
"The Jersey Jumper," Sam Patch

"Hail to the hero, Samuel Patch, who knows not any equal; In jumping, Sam can find no match, Among ten million people"

The mark of a hero in 1829 was to be eulogized in poetry, and Sam Patch of Paterson knew the satisfaction of being thus noted in the *Saturday Evening Post*. The more he jumped, the higher he rose. And he would also become known by a renowned catchphrase: "Some things can be done as well as others." Yet America's fascination and captivation with showmen and celebrities started very early, as Sam Patch's story illustrates.

It all began in Pawtucket, Rhode Island, when a spinning mill was opened there in 1789, powered by the fifty-foot falls of the Blackstone River. But it was a financial failure until Samuel Slatter arrived from England in 1790, his head filled with patents and high-tech systems England was trying to keep secret. He assured the owners that if they made him a partner and he didn't make a yarn as good as was made in England, he would throw the "whole of what [he] attempted over the bridge." Regardless of whether Samuel Slatter was offering his actual suicide or merely an alliterative death, it is clear there was something about that bridge fifty feet above the falls that inspired men to envision a leap of faith. Sam Slatter did make good yarn, and his mill became ground zero for America's Industrial Revolution.

At the age of six, Sam Patch was abandoned by his alcoholic father, Greenleaf Patch. At seven, he joined his siblings in tending to the spinning mules in Mr. Slatter's mill. Patch spent his youth there as a "doffer," which required him to scuttle between the working looms, replacing bobbins and re-directing errant shuttles. Work began at five o'clock in the morning and did not end until half past seven, six days a week. The usually exhausted Sam was lucky not to be disabled by the unshielded whirling belts and flying equipment just inches from his head and hands. He was eventually promoted to weaver and received a weekly salary of $2.50. In the summer, while on his half-hour lunch breaks, Sam often threw himself off the bridge into the cool water of the tidal pool at the foot of the falls. Each plunge was a brief moment of weightless freedom, an escape from his pitiless existence. He would also sometimes make dramatic leaps off the roof of a four-story building into the aerated churning water below the falls.

In 1824, the mill owners in Pawtucket demanded that workers accept a 25 percent cut in wages and scale back their lunch breaks to fifteen minutes. Sam Patch, having fourteen years' experience in the mills, helped to organize the first workers' strike in America. Forced to back down, the owners removed as many of the troublemakers as they could, and Sam was forced to leave Pawtucket.

By 1826, he had found work as a loom supervisor at the Hamilton Cotton Mill in Paterson, powered by the high falls at the Passaic River. He was now an abusive alcoholic himself, an angry and not particularly admirable man who was known to box the ears of the young doffers working under him.

While in Paterson, he started a cotton mill of his own and partnered with an Englishman, but the business failed when the partner ran off with the funds. Twice, Patch tried to run a mill of his own, and twice, his addiction to whiskey spoiled his chances. Some believed Patch began to jump for purses of money. Until 1827, only those close to his cotton-spinning machine in the Paterson mill heard his boasts. Shortly after his twentieth birthday, he boasted that he would jump over the Passaic Falls.

On the western shore of the Passaic Falls, there was an island of escape called the Forest Garden that was often used by working-class people for picnics. On August 14, 1827, it was purchased by industrialist Timothy Crane, owner of a grist- and sawmill on Van Houten Street in Paterson. Crane then transformed the idyllic spot into a private park, complete with scenic walks, an upscale restaurant-tavern named The Cottage on the Cliff, beer gardens and well-manicured versions of nature to cater to the wealthier residents of the area. He began building a new wooden, pre-constructed bridge over the Passaic Falls to his garden to restrict access to only those who could afford the two-cent toll to keep out the riff-raff. The working-class locals, enraged at his exclusionary toll, began to beat up the children he employed as toll takers. It was a big event. The factories were even closed that day.

The bridge was assembled on September 20, 1827. Crane staged a grand celebration as the bridge was pushed out across the Falls. A small crowd gathered to cheer the endeavor. More disenfranchised millworkers stood about watching their picnic grounds being claimed by a wealthy overlord. Who then should step out on a rock outcrop above the Falls but the weaver Sam Patch, proudly dressed in his white uniform. The police who were patrolling the crowd were horrified. The drunken Sam had been locked up in a basement all morning to keep him away from the ceremony. Somehow, he had escaped, and they were worried he might start a riot. Sam shouted out to the crowd, but he didn't call for violence. He simply announced that Timothy Crane had indeed done great things, but now Sam Patch would do great things as well. William Brown, a witness, remembered that Patch appeared composed and recalled that he took a couple of small stones in his hands, went to the brink of the cliff and dropped them off, one after the other. He watched where they struck the water down below. Then he walked back a few yards, turned and took a little run to the brink of the cliff and jumped off, clearing the rocks by about ten feet. He went down feet first, but with his body inclining considerably to one side, and in this shape, he struck the water and disappeared. For several seconds, the crowd was convinced he had died on impact. But then, a few seconds later, his head bobbed up and he began paddling for shore at a point downstream. Patch had not only jumped before the cheering crowd but also retrieved a large piece of wood that had drifted as the bridge was being put in place. Sam Patch had his day of fame.

After Sam made his famous leap at the Passaic Falls in September, explanations for his conduct were offered. Accounts differ as to the circumstances of Sam's jump. There were reports that he had planned the leap as a workingman's protest against Crane and the conversion of the picnic areas to a private playground for the rich. Others said that Sam was drunk and decided to jump at the spur of the moment to retrieve a roller that had fallen from the apparatus being used to move the bridge into place. Some even said that a jilted Sam had jumped for love, since it was said that Patch was unlucky in love and had drunkenly attempted to put an end to his miseries. It had been nothing but a failed suicide leap. But Patch stated that simply wasn't true: "I am perfectly sober and in possession of my proper faculties and leaping waterfalls is nothing more than an art which I have knowledge of and courage to perform, an art which I have practiced from my youth."

The citizens of Paterson were impressed with his feat, except of course for Timothy Crane, who felt the drunken lout had stolen his thunder. And he had. Crane tried to get it back on Wednesday, July 4, 1828. That evening he was staging a ticket-only display of fireworks to be held above his exclusive Forest Garden, on the far side of the Passaic Falls, "where the refinements of taste and art combined with the varied and romantic beauties of nature to afford pleasure and satisfaction to the numerous company present." This was the first Fourth of July celebration in Paterson to feature such advertised commercial attractions. The previous year, the fireworks display had been paid for by the municipality and had been free to all. But that afternoon, Sam Patch did it again. He leaped off the same rock before an estimated crowd of five thousand and plunged seventy-seven feet into the water. The newspapers reported these words of wisdom from the "Yankee Jumper": "Some things can be done as well as others."

People couldn't stop talking about Patch's amazing death-defying leaps. Once again, he had upstaged Timothy Crane. Sam Patch did it a third time on July 19. He was promised $15,000 for his daring exploit. He jumped the Passaic Falls again for a crowd estimated to be between six thousand and ten thousand people—a crowd considerably larger than the resident population of Paterson.

Then on August 11, 1828, in Hoboken, he leapt one hundred feet from the mast of a sloop and splashed down into the Hudson River. The paying crowd only numbered five hundred in Hoboken, but the New York newspapers had begun to take notice of "Patch, the New Jersey Jumper," the working stiff risking his life to make a living.

Sam Patch's next opportunity arrived in the form of an invitation from businessmen in the city of Niagara Falls, New York. Sam Patch had become the Evel Knievel of his day, entertaining crowds by jumping feet first into rivers adjoining cliffs. After each successful jump, he would make a second, more difficult jump from a platform twenty-five feet higher than the cliff. At Niagara Falls, a cave had recently been discovered; it protruded from beneath an outcrop beneath Goat Island, which divided the American Falls. Seizing upon the cave's official opening for tourists on October 5, a dive by the Yankee Leaper off a 125-foot ladder against the backdrop of the falls at the exit of the Cave of Winds seemed the perfect attraction for the opening. And they only had to pay him seventy-five dollars for the stunt. The only problem was Sam was now suffering from bouts of delirium tremens and missed his jump-off date. He apologized to those who had not yet left town and assured them he would venture a small leap of eighty feet that Wednesday, if only to convince those who had stayed to see that he was the true Sam Patch and to show that "some things could be done as well as others." A crowd of ten thousand showed up to see if he would make the

Illustration from a story entitled "Sam Patch's Last Leap." It said that his last leap was so deep, he wound up in the South Seas. *Courtesy of the author's collection.*

leap, and he did not disappoint. He came down feet first into the whirling white eddies below the Falls and then did it again on Saturday, October 17, this time in a pouring rainstorm. As he climbed out of the river after the last jump, Sam greeted the adoring crowd: "There's no mistake in Sam Patch!" He was a hit. He had become the first daredevil to jump off Niagara Falls and live. He jumped several more times from Goat Island into the Niagara River before he met his end in Rochester.

After his Niagara Falls leap, Sam Patch was immediately booked to leap from atop the ninety-nine-foot-high falls of the Genesee River. Patch hurried to Rochester, New York, the midpoint of the Erie Canal, to fulfill the jumping commitment before the river froze. For practice, he jumped the Upper Genesee Cataract privately, witnessed by his companion, Joe Cockrane. On November 13, a Friday, handbills distributed at the Genesee Falls in New York State advertised "Leaping Sam Patch's Last Leap." This was because Patch intended it to be his final jump in the United States before moving on to perform in Europe. The sign, however, would prove more prophetic than intended and become more final than the loophole Sam had planned.

On November 6, Sam showed up a little before 3:00 p.m., accompanied by his pet bear cub, to a crowd of between six thousand and ten thousand people. He started the show by throwing his bear cub over the side, and when the crowd witnessed the bear climbing to shore, Sam then jumped as well. He lived once again, but the collections from the crowd were so disappointing that he deliberately picked the unlucky date of Friday the thirteenth to repeat the feat for a larger audience.

A crowd of about eight thousand to ten thousand people was waiting at the ninety-foot Upper Genesee Cataract at 1:00 p.m. As was his usual custom, Sam drank liquor prior to jumping. It may have played a role in his death. He appeared before the crowd wearing white pantaloons with a red sash around his waist and a jacket. He made his jump again, this time from a twenty-five-foot tower he had built at the brink of the precipice. It was publicized as "Higher Yet! Sam's Last Drop!" And it was. As he climbed the ladder, some would later say that Sam Patch staggered a bit. He had taken at least a single glass of brandy before climbing that ladder. Once atop the tower, he waved his hand to the crowd and called out:

> *Napoleon was a great man and a great general. He conquered armies and he conquered nations, but he couldn't jump the Genesee Falls. Wellington was a great man and a great soldier. He conquered armies and he conquered nations, but he couldn't jump the Genesee Falls. That was left for me to do, and I can do it, and will.*

Patch then made his jump, straight as an arrow, but he was witnessed to go limp as he plummeted to the water at the base of the falls. He flailed

Sam Patch was famous all over as the first daredevil in America. He jumped the falls at Pawtucket on the Blackstone and jumped right into American legend. *Courtesy of the author's collection.*

his arms on the way down. His legs were not together, and he began to lean. He entered the water at an angle, and there was a noisy impact. There was a huge splash, and he was not seen after his jump. When the water calmed, there was nothing. Patch failed to resurface. The crowd hung in there, anticipating a dramatic resurfacing that never came. Even months later there was speculation that Sam had hidden in a cave beneath the falls and then run off. Some believed that he'd pop up later and it would turn out to be a publicity stunt. There were reports of Sam Patch sightings all over the country, but Sam Patch was no more, dead before the age of thirty.

They dragged the river but did not find his remains. Speculation that Patch had hidden in a cave at the base of the falls was dispelled months later when the river began to thaw. On March 17, 1830, Silas Hudson, a farmer, recovered Patch's frozen corpse at Charlotte, near the mouth of the river. Silas broke the ice of the Genesee River five miles downstream near the river's joining with Lake Ontario. As Silas's horses drank the cold water, he claimed he saw a body under the ice, jammed against the shore. They identified the corpse by the black scarf around his neck and the frozen features. Sam Patch

"The crowd, with fingers in their mouths/Turned homeward one by one/And oft with sheepish looks they said/ Poor Sam's last job is done." *From the poem "Biography of Sam Patch" by Major Jack Downing. Courtesy of the author's collection.*

was buried near where they found the body in the Charlotte Cemetery, on River Street in Rochester. The grave was initially marked with a wooden board that read SAM PATCH—SUCH IS FAME. The plank was initially carried in local parades and then returned to his grave after each occasion before it was eventually stolen. Later, a proper gravestone marker was paid for by donations collected by students of Charlotte High School in the late 1940s, but the birth date was wrong. It also said Patch had leaped to his death over the upper falls, as if he had committed suicide. But then "suicide" implies Sam Patch had a choice. Today, there is a bronze plaque on a cemetery stone marking the grave.

Sam Patch's life and death inspired multiple authors of the era, including Nathaniel Hawthorne and Herman Melville. Additionally, he was commemorated in poems, including one by William Carlos Williams. He is also mentioned in a line in *Walden* by Henry David Thoreau. There was even a play written about Sam's life, entitled *Sam Patch: The Yankee Jumper*. Immediately following Sam's final leap, stories spread of a phantom figure seen leaping from the edge of the falls on some nights. The stories of Sam's ghost have faded over time.

Chapter Nine

The Great Falls Festival

In 1971, a committee called the Project Committee was appointed by Paterson's mayor, Lawrence F. Kramer, to renovate the Falls and vicinity. As a result of its efforts, a new steel bridge replaced the skeleton of the structure that had been closed since April 1917. The area of the Falls was turned into an attractive park. A new lighting system was installed, and much of the river's debris was removed. The completion of the project was celebrated by a Great Falls Festival, which was held during the week beginning September 1, 1971.

The first Great Falls Festival was spearheaded and planned in a matter of weeks by Mary Ellen Kramer and was held September 4, 1971, Labor Day weekend. It was Mayor Kramer's way of celebrating the opening of the park, bringing the people of Paterson together to have a good time, honoring the city's National Historic District and celebrating the labor movement. The event even included the return of a daredevil act over the Falls, and the governor, William T. Cahill, was on hand to cut the ribbon at the bridge and to pay tribute to the city and those who had worked on the project. Charles Ward, who had closed the old bridge in 1917, opened the gates to the new one. Daily circus acts, band concerts, exhibits and tours entertained visitors. The festival was a tremendous success and brought tens of thousands of citizens to the area. All proceeds went to the Great Falls Development Corporation for the restoration of the Falls. It was viewed as an act of dedication to the Falls, the city and its future.

The festival ran continuously until 1980. Each year, the festival featured an aerial act, including appearances by perhaps the two most famous

Jill Dente works at a booth at the first Great Falls Festival, Labor Day Weekend in 1971. That festival drew a crowd of thousands to the area. *Photo by Marcia Dente.*

practitioners in the world, Philippe Petit and Karl Wallenda. Daredevils can best be summarized as persons who wish to take conscious risks with their lives with the emphasis on survival. Today, the art of being a daredevil has become so sophisticated that chances of survival are almost predictable. However, it can't compare to those of yesterday because of changing conditions and technological innovation. Most of the contraptions the daredevils used then usually had a nickname or had a statement printed on them. The daredevils pursued fame and fortune or notoriety, yet none has ever become rich nor had lasting fame.

The first daredevil to walk across the Falls on a tightrope was Mons de Lave in 1860. He was engaged by Albert van Voorhies, who leased and operated the Cottage on the Cliff in the summer months for visitors who came to see the Great Falls. De Lave, who achieved some notoriety by crossing Genesee Falls on a tightrope, gave a similar exhibition at the Falls, which was witnessed by thousands of people, some of whom had paid a quarter and some of whom had not. The place selected for the "Grand Ascension" as it was called was below the Falls, where the river passes through a gorge at least six hundred feet wide and nearly two hundred feet deep. The hemp rope that was used stretched across this chasm from the Cottage on the Cliff to the top of Morris' Mountain. De Lave made his appearance shortly after four o'clock on a day in May and mounted the ladder leading to the starting point on the rope. He was greeted with cheers, which he acknowledged by taking off his cap and waving to the multitude below. He poised himself on the rope, pole in hand, and began the walk, keeping time with the band, which played "Hail Columbia." Each person held his breath as the little Frenchman walked rapidly forward over the dizzying height. The rope swayed slightly under his step, but there was no interruption in his progress, and in about four minutes, he reached the opposite side. Prolonged cheers greeted him from both sides, and de Lave passed his cap among the crowd afterward, accepting whatever the throng threw into it, which he said was a handsome sum. Fifteen minutes later, he appeared again, waving a signal to the band, which struck up "La Marseillaise," and he began his return. In the middle of the rope, he stopped, lay down on his back, rose, sat astride the rope and then repeated these feats several times. Spectators from both sides were cheering. After his walk, he appeared winded and was perspiring profusely. But he gave notice that he would repeat his walk across the Falls the following Monday.

Mons de Lave, a Frenchman, was the first daredevil to attempt to walk across the Falls. His feat was witnessed by thousands of onlookers below and above the chasm. *Courtesy of the Paterson Museum.*

He walked with peach baskets tied to his feet, stood on his head and performed other terrific feats of daring.

Harry Leslie, an English professional tightrope walker known as "the American Blondin," crossed the Passaic River on August 5, 1879, just below the Falls in the presence of several thousand spectators below and on both sides of the chasm. The river at that point ran through the deep chasm, and some five hundred feet of rope was needed to span it. The rope was made of silk and was carried across the river at a height of sixty to one hundred feet above the water, and it stretched from the pine tree (made famous as the spot where Sam Patch made his leap and where a Mrs. Hooper Cummings, a visitor to the Falls, fell in 1812) to Morris' Mountain. Sometime after 3:00 p.m., Leslie stepped out on the rope, danced lightly across the fearful chasm, paused midway to kneel on the rope, hung from it by his toes and performed many similar feats. After crossing, he returned immediately from Morris' Mountain back to his starting point.

Thirty thousand people held their breaths while another French aerialist, Philippe Petit, crossed a cable strung 100 feet above the swirling waters of the chasm at the Great Falls. Twenty-five-year-old Petit had recently startled the world with his walk between Manhattan's World Trade Center Towers in August 1974. The walk across the Falls' chasm lasted a precarious eight and a half minutes, each step a dizzying one on the 208-foot-long wire. A sign reading "PATERSON" was suspended on the bridge, and the Great Falls itself was flashed across TV screens the world over. More than forty-five thousand people attended the Labor Day weekend celebration that year. After the act, Petit told Paterson's mayor, Lawrence Kramer, that he was an artist and not a circus entertainer. The mayor discovered that for himself as he fed Petit gallons of orange juice during the long weekend and stood by while the setup was rigged and rerigged following a night of torrential rain.

No one will ever forget the sight of seventy-year-old Karl Wallenda standing precariously over the waterfall in a billowy white shirt and black pants on a one-inch-wide steel cable without a net beneath him. The highlight of the 1975 Great Falls Festival, literally and figuratively, was watching as Wallenda clenched a fifteen-foot-long balancing pole in his hands and jittered seismically. Wallenda squinted into huge spotlights that were focused on him and carefully placed one slippered foot in front of the other as he moved across the wire, enveloped by mist and spray from the Great Falls. As Wallenda took his final steps on the wire and stepped off onto the western cliff of the Falls, he fell into the arms of a relieved and jubilant crowd of Patersonians. People reached out to hug him and shake

Another Frenchman, Philippe Petit, walked across the Falls in a high-wire act for more than thirty thousand people over a four-day period in 1974. *Courtesy of the Paterson Museum. Photo by Michael Spozarsky.*

his hand and then stared at their hands and arms in amazement: Wallenda was soaking wet.

Three years later, Wallenda was walking across a wire strung between two high-rise buildings in San Juan, Puerto Rico, when he was knocked over by a gust of wind and fell to his death.

There have been many high-wire acts over the years at the Falls, as well as helicopter and motorcycle acts, aerial acrobat and a human cannonball that was shot head first with only a net above the Falls to catch him. In 1991, Tino Wallenda, grandson of Karl Wallenda, crossed a cable stretched one hundred feet over the churning waters and lifted his arms to the cheers of over one thousand spectators when he finished his act. In 1992, on the last day of the festival, John Lemoine and his motorcycle daredevils, on the spur of the moment, took Paterson's mayor, Bill Pascrell Jr., across the gorge and

"The only place I feel alive is on the wire," said Karl Wallenda about his high-wire act. Wallenda is the patriarch of his well-known family of circus performers. *Courtesy of the Paterson Special Events Office.*

back, strapping him to a trapeze bar attached below the racing motorcycle. Despite warnings not to look down, Pascrell wouldn't miss the dizzying view. The motorcycle abruptly shot backward across the river and set him back safely into the waiting arms of relieved handlers.

The Great Falls Festival had an eleven-year run before a change in the administration. The festival was canceled in 1981, the administration citing the strain on the city's financial resources. The last festival had drawn more than 100,000 people over Labor Day Weekend. But former mayor William J. Pascrell Jr., currently the congressional representative in the Eighth Congressional District and one of the members of the original Great Falls Committee, brought back the festival to cheering crowds in 1991. He resurrected a scaled-down version of what took place in previous years' celebrations. The Great Falls Environmental Festival was held over the Fourth of July holiday (as it had been held in the 1800s) rather than Labor Day weekend. For several years, the event combined the historic attraction with the commemoration of Independence Day. On September 3, 1994, the 100th anniversary of the Labor Day movement, the Great Falls Festival was returned to its Labor Day weekend celebration and included a Labor Day parade that began at the American Labor Museum in Haledon (the Botto House National Landmark, the site of refuge for the silk strikers of 1913) and ended at the Great Falls.

Chapter Ten
National Park Designation

The roots of the park effort date to the 1960s, and the Great Falls and the surrounding historic mill area have been the focus of preservation efforts. In the 1960s, well-developed plans by state and city officials to improve automobile access into downtown Paterson (and thereby stimulate much-needed urban renewal) aroused the concerns of local citizens, who feared the impact of such a scheme on the city's unique historic industrial fabric. With the state already beginning to acquire the right of way through the city, a small, informed preservation group was formed under the leadership of the late Mary Ellen Kramer, wife of Lawrence F. Kramer, who was mayor at the time. Mary Ellen realized the dramatic effect on the city that this major highway project would have. The state wanted to run Route 20 right through the downtown area and tear down dozens of historic buildings, but she was one of the key people who helped to stop that project and the driving force behind a newly formed citizens group, the Great Falls Committee. In 1971, the Great Falls Committee was incorporated as the Great Falls Preservation and Development Corporation, a nonprofit public membership organization that sought to guide the redevelopment of Paterson's historic downtown and concurrently pressed for the preservation and restoration of downtown Paterson, including the raceway system and many of the mills. Under Mary Ellen Kramer's leadership, the committee and its successor played a critical role in the creation of the Great Falls Park (now Mary Ellen Kramer Park), the establishment of the Great Falls Festival and the restoration and reactivation of the hydroelectric plant at the foot of the Great Falls. Beginning in the 1970s, the same citizens'

Former speaker of the house Nancy Pelosi joined Congressman Pascrell at the Great Falls prior to the historical park designation. *Courtesy of the author's collection.*

movement sought the involvement of the National Park Service and the Smithsonian Institution in its efforts to heighten awareness of Paterson's industrial heritage. The committee, however, only enjoyed a few short-lived successes in its early years.

Overlook Park was created and dedicated on April 16, 1964, during the administration of the late mayor, Frank X. Graves Jr., and was spearheaded by Paterson newspaper publisher Harry B. Haines. The park's creation marked the city's recognition of its historic heritage. It was originally called Harry B. Haines Memorial Park.

In 1967, the National Park Service designated the Great Falls and ten acres of Garret Mountain as a National Natural Landmark. According to the National Park Service, the area "provides an excellent illustration of the jointed basaltic lava flow that began a period of extrusion and intrusion throughout eastern North America in the early Mesozoic, influencing present day landforms in this region." In 1977, the Great Falls Preservation and Development Corporation successfully landed the waterfall vicinity on the National Register of Historic Places. On the momentum of those victories and with hopes of establishing the country's first urban national park in Paterson, the group persuaded the Department of the Interior to perform a major survey of the area. The department's National Park Service was considering Paterson; Lowell, Massachusetts; and Troy, New York. Ultimately, Paterson lost out, and Lowell received federal designation in 1978, which brought an infusion of cash and helped spark that city's recovery. But it meant Lowell officials had to cede some control to Washington. There's always been some hesitation in Paterson about being controlled from the outside. Even after the great fire in 1902 that destroyed the downtown area, there were people trying to donate money from the outside, but Paterson wouldn't accept it.

The Great Falls Preservation and Development Corporation faced political opposition from those who wanted to see the historic district entirely redeveloped, back when federal urban renewal money seemed to be flowing more freely. In the early 1990s, U.S. senator Frank Lautenberg secured a $4.1 million grant to preserve the city's historic buildings and clean up the raceway system. Only a handful of structures were renovated, including the Rogers Locomotive Works, which today houses the Paterson Museum, and the Essex and Phoenix Mills, which both function as affordable artists' housing and office space.

On April 17, 1970, the Great Falls/SUM district was added to the National Register of Historic Places. The district encompassed 1,080 acres, forty buildings and two structures. On May 27, 1971, the area was added to the New Jersey Register of Historic Places.

The Great Falls Preservation and Development Corporation sponsored a "Winter Walk of Paterson and the Great Falls" on February 15, 1976. The Falls area had always attracted tourists, and this new publicity brought people from all over the world to visit and participate in scheduled and

A young Henry Papale and his mother, Roberta, visiting from Canada, view the Falls from Mary Ellen Park. The park provides paths around the area that was previously off-limits to visitors. *Photo by Marcia Dente.*

unscheduled events. On May 15, 1977, a twelve-mile bike tour featuring the Historic District was held.

The district also served as the backdrop of a feature-length film written and directed by two residents of the Eastside area. The Ryle House was the "set," and there were many shots of the Historic District, including a murder scene from the interior of the unrestored Rogers Building. Titled *Communion*, the film premiered in Paterson on November 13, 1976, at the original Fabian Theater, and the proceeds from this gala went to the Great Falls Preservation and Development Corporation. Ultimately released as *Alice Sweet Alice*, the film is most notable for serving as twelve-year-old Brooke Shields's debut.

On the rainy afternoon of June 6, 1976, while standing on a dais at a dedication ceremony, President Gerald R. Ford made a bicentennial speech in Paterson with the spectacular natural beauty of the Great Falls in the background. During the speech, the sun came out, rays peeked through a dense downpour and Ford, who was standing directly in front of the spray, spoke against its roar. In his opening remarks, he referred to Mayor Kramer, New Jersey governor Brendan T. Byrne, state senator Pete Williams, several congressmen in attendance and Edward B. Haines, executive editor of the *Paterson Evening News*. Ford proclaimed:

> *Today the Great Falls of the Passaic, Society for Establishing Useful Manufactures' Historic District takes its place alongside such sites as Mount Vernon, Monticello, as a national historic landmark. That makes this a very important day in Paterson—not just for Paterson, but for the entire United States. All America can look at this landmark and consider what it really represents. Great Falls is joining twenty-one other landmarks which were the settings for important facts and important acts in the drama of our nation's industrial development. But this site has a very particular significance within that very select group. The sheer number of different industries that were established here, make this site very unique.*
>
> *Great Falls represents the first attempt within the United States to harness the entire power of a major river. Others in Hamilton's day, a good many years ago, came here to Paterson to admire the Great Falls. They saw its beauty as its only natural resource. But Hamilton found more than just the beauty of the Great Falls which delighted his eyes. When Hamilton first saw the Falls, as an aide to General Washington, they fired Hamilton's imagination. He saw them as a power supply with raw materials, abundant labor, and ready markets nearby.*

Before Paterson was much more than a name on a piece of paper, Hamilton saw it as the cradle of American industrial might. A lot of water has flowed over the Great Falls since those days. Paterson did become a cradle which nurtured such American industrial products as Rogers's locomotive, the Colt revolver, the Holland submarine and the Curtiss-Wright aircraft.

Today, we are making the historic district a monument to the genius and to the efforts of the workers, the engineers, and the businessmen who turned Hamilton's vision of an industrial, powerful nation into a reality. We can see the Great Falls as a symbol of the industrial might which helps to make America the most powerful nation in the world, a nation that each and every one of us can be very proud of in 1976, our Bicentennial Year.

We can see it as a symbol of industrial democracy, which makes a vast array of material goods available to all of our people. The ancient energy of the Great Falls themselves and the new energy of the citizens of Paterson can propel this area into a new age, even greater than that foreseen more than 200 years ago by Alexander Hamilton. Those same natural resources, the American earth and the American people, will ensure that his vision of a free and prosperous nation will remain a reality for a long, long time.

The 1976 Bicentennial Committee chaired by Paterson council president Frank X. Graves Jr. planned for the arrival of President Gerald Ford at the Great Falls. *Photo by Marcia Dente.*

That day, with sheets of rain that swelled the Passaic River and water thundering down the seventy-seven-foot precipice behind him, President Ford designated the Great Falls, as well as the Society for Establishing Useful Manufactures' historic district and its 119 acres, as a National Historic Landmark. The speech was made with sixty thousand residents in attendance.

On May 20, 1977, the American Society of Civil Engineers (AMSCE) and the American Society of Mechanical Engineers (ASME) designated the Great Falls Raceway and Power System as a National Historic Civil and Mechanical Engineering Landmark. A plaque was dedicated and placed at the upper raceway commemorating the occasion. The plaque reads, "The Great Falls Raceway and Power System at Paterson is the basis of the oldest American community integrating water power, industrial development and urban planning." The date of this dedication coincided with the fiftieth anniversary of Charles Lindbergh's crossing of the Atlantic Ocean in an airplane powered by a Wright Aeronautical engine made in Paterson.

Standing between the second-largest waterfall east of the Mississippi River and more than one hundred residents, former governor James E. McGreevey announced on October 21, 2004, that the Great Falls was officially adopted as an urban state park in New Jersey.

At a press conference at a site overlooking the Falls, McGreevey, stated:

> *Today is an important day in which we celebrate not only God's natural beauty, but the people of Paterson, the ones who worked in these mills and*

Former governor James E. McGreevey made his formal adoption speech of the Great Falls State Park on October 21, 2004, at the Great Falls. *Courtesy of the author's collection.*

> *who walked these streets. Some of our most significant historical areas are in our cities and our children in cities need open space and green space as much as any other child in New Jersey. Our parks and open spaces are the pride and joy of New Jersey. Parks create a sense of place, provide recreational and educational experiences, protect the environment and boost the economy. A world of opportunities and advantages rests in these new parks, which benefit not just our urban centers but all of New Jersey.*

The current plan for the Great Falls Park was developed by Field Operations, an internationally known landscape architecture and design firm. Its plan is to create intimate "green" spaces and "living rooms" out of industrial urban blight without erasing any sign of the buildings' histories. The final draft plan will transform the Allied Textile Printing site, a seven-acre parcel of land owned by the City of Paterson, into a walking area and an archaeologically significant place where residents and visitors can learn about Paterson's powerful industrial legacy and its role in paving America's economic path to independence.

For the past ten years, Paterson's representatives in Congress have put forward a bill every year to make the Falls a national park, but none of those bills were ever brought to the floor for a vote until 2007, when one of the measures, sponsored by Congressman Pascrell, finally passed the House of Representatives and then the United States Senate. This vote was the culmination of decades-long efforts by local officials to transform the historical site into a national entity and the realization of a dream that Congressman Pascrell has had since first becoming a member of Congress in 1996.

At a ceremony in the East Room of the White House on Monday, March 30, 2009, President Barack Obama signed into law the Paterson Great Falls National Historical Park Act. The Paterson Act is part of an Omnibus Parks and Preservation Bill that the president called "one of the most important pieces of natural resource legislation in decades."

> *The passage of this Act honors and preserves Paterson's past. As Americans, we possess few blessings greater than the vast and varied landscapes that stretch the breadth of our continent. Our lands have always provided great bounty—food and shelter for the first American, for settlers and pioneers; the raw materials that grew our industry; the energy that powers our economy. What these gifts require in return is our wise and responsible stewardship. As our greatest conservationist President Teddy Roosevelt put*

On March 30, 2009, in the East Room of the White House, President Barack Obama signed the Omnibus Public Lands Management Act. *Courtesy of the author's collection.*

> *it almost a century ago, "I recognize the right and duty of this generation to develop and use the natural resources of our land; but I do not recognize the right to waste them, or to rob, by wasteful use, the generations that come after us." That's the spirit behind the bipartisan legislation I'm signing today—legislation among the most important in decades to protect, preserve and pass down our nation's most treasured landscapes to future generations.*

When the proposal became law, this site became part of a network that includes great wildernesses, such as the Grand Canyon and Yosemite, and urban areas, such as Lowell National Historical Park in Massachusetts.

On April 15, 2009, city residents and activists crowded under a white tent set up opposite the Falls and praised the teamwork of the many involved in gaining national park status, what they once felt was a seemingly impossible goal. Congressman Pascrell said the Great Falls represents the history of the people of Paterson. "We don't forget where we came from," he said. "These are our roots. This is our Plymouth Rock." According to New Jersey senator Bob Menendez, the Great Falls is just as significant to the history of the American people as some larger and more well-known parks,

After signing this historic legislation, the president called the act the most important piece of natural resources legislation in decades. *Courtesy of the author's collection.*

such as Yellowstone National Park. But our national park system isn't just about giant spaces out west. It's also about preserving places of national significance here in New Jersey, which sustains our history and provides recreation for everyone.

On November 7, 2011, four years after the United States House of Representatives voted to create the Paterson Great Falls National Historical Park, secretary of the interior Ken Salazar and Paterson mayor Jeffery Jones signed a historic agreement that paves the way for establishing the new national park. U.S. senators Frank Lautenberg and Robert Menendez, Representative Bill Pascrell Jr., National Park Service director Jon Jarvis and Darren Boch, who was recently named superintendent designate for the Paterson Great Falls National Historical Park, all joined Secretary Salazar and Mayor Jones. The agreement Secretary Salazar and Mayor Jones signed transferred property and established easements that will fulfill the requirements of the law and authorize the establishment of the new national park signed by President Obama in March 2009.

According to Secretary Salazar, "Paterson and its Falls played an integral role in the industrial growth of our nation and in the lives of immigrants who labored in the mills and ultimately joined unions to seek better working

On November 7, 2011, the new sign that now welcomes visitors to the Paterson Great Falls National Historical Park was unveiled after the park was dedicated. *Photo by Marcia Dente.*

conditions and pay. By establishing this park, we not only tell the story of Paterson, but we also contribute to the economic growth of the city today by attracting visitors and supporting jobs in local communities."

Senator Bob Menendez said, "From the Great Falls, through the raceways and water wheels along the Passaic, flowed the blood, sweat, and tears of the men and women who powered the Industrial Revolution and made this nation great. The park's history is now part of the story of America."

Senator Frank Lautenberg poked fun at himself for being the oldest member of the U.S. Senate and joked that he was with Alexander Hamilton in 1778, when he looked out at the Falls and envisioned America harnessing the water to become a great industrial power.

> *Years later, as the country's first Secretary of the Treasury, Hamilton selected the site to become the nation's first planned industrial city. This is an historic time for the Great Falls and my home city of Paterson. With this designation, the Great Falls is America's newest national historical park, and one of our nation's most beautiful and historic landmarks and will finally get the recognition it deserves. This new park will showcase the majesty of the Falls and encourage more tourists, families, artists, students and businesses to come to Paterson and help strengthen this great city. The dedication that Secretary Salazar made today is much like the Great Falls—a testimony to the past work that has brought us to where we are today and a promise of the rewards and achievements we have yet to realize. Today marks a momentous occasion that will serve as the starting gun for the establishment of Paterson Great Falls National Historical Park. I truly believe that our prized natural and historic resource will one day be the crown jewel of the National Park System.*

Mayor Jeffery Jones shared the following:

> *Just as our Founding Father Alexander Hamilton envisioned using the power of our magnificent Great Falls to help create a self-sufficient new nation, I am envisioning the many possibilities that will present themselves for our city and our citizens. Paterson Great Falls National Historical Park not only heralds our strong past, but more importantly offers Paterson the potential to take a step back in time with our feet firmly planted in the now. Utilizing the examples of the historic use of the waterfall, and the surrounding areas, we can clearly visualize embracing alternative energy sources in the form of hydroelectric power, wind power and solar power to re-emerge as a leader in the American renewable energy revolution.*

Congressman Pascrell, who once was mayor of Paterson and who spearheaded this effort, finally succeeded with the Paterson Great Falls National Historical Park Act of 2000, HR280. Every member of the New Jersey congressional delegation cosponsored this bipartisan legislation. The language from that bill was incorporated into the larger Omnibus Public Lands Management Act of 2009, HR146. "We have finally done it," said

Congressman Bill Pascrell Jr. stands with the Great Falls as a backdrop after the official designation of the Paterson Great Falls National Historical Park. *Courtesy of William J. Pascrell Jr.*

Congressman Pascrell. "Paterson will finally be recognized for the seminal role it has played in shaping American history…This place is a hidden jewel; the history is rich here and this city really deserves this kind of recognition. There's no place like it in the United States."

No other natural wonder in America has played such an important role in our nation's historic quest for freedom and prosperity. Alexander Hamilton conceived and implemented a plan to harness the force of water to power the new industries that would secure our economic independence. Today, the Great Falls, a spectacular natural wonder of the Passaic River, is the only National Historic District that includes both a national natural resource and a National Historic Landmark. Scholars have concluded that Pierre L'Enfant's innovative water-power system in Paterson and many of the factories built later in the town constitute the finest remaining collection of engineering and architectural structures that represent each stage of America's evolvement from a weak agrarian society to a leader in the global economy.

This small poster was distributed by the New Jersey Community Development Corporation, headquartered in Paterson. It is committed to revitalizing the community. *Courtesy of the New Jersey Community Development Corp.*

Heavy rainfalls give the Falls a chance to show its glory to visitor Bill Hansen during his first visit to the Great Falls. *Photo by Bill Harris.*

The story of Alexander Hamilton and American industry is not just a story of the state of New Jersey but also the story of our nation. Recognizing this, Congress and President Obama have seen the need to preserve and protect the Great Falls National Historic District. It is more than an outstanding scenic attraction in a densely developed urban area. Paterson, the Great Falls and the land surrounding—all a birthplace of the Industrial Revolution in America—are national historical treasures, and with the new legislation, Americans from all over the country will be coming to visit. And the mighty Great Falls awaits them as William Carlos Williams writes in his epic poem: "The river comes pouring in above the city and crashes from the edge of the gorge in a recoil of spray and rainbow mists."

BIBLIOGRAPHY

Adler, Dennis. *Colt Single Action: From Patersons to Peacemakers*. Edison, NJ: Chartwell Books, 2007.

Archdeacon, Peter. *A Sketch of the Great Falls of Paterson*. New Jersey: 1845.

Avignone, June, ed. *Downtown Paterson*. Charleston, SC: Arcadia Publishing, 1999.

Bontemps, Arna, and Jack Conroy. *Sam Patch: The High, Wide and Handsome Jumper*. Boston: Houghton Mifflin, 1951.

Brydon, Norman F. *The Passaic River: Past, Present Future*. New Brunswick, NJ: Rutgers University Press, 1974.

Capo, Fran. *It Happened in New Jersey*. Guilford, CT: Globe Pequot Press, 2004.

Chernow, Ron. *Alexander Hamilton*. New York: Penguin Books, 2004.

Cunningham, John T. *This Is New Jersey*. New Brunswick, NJ: Rutgers University Press, 2001.

Dente, Marcia. *Great Falls of Paterson*. Charleston, SC: Arcadia Publishing, 2010.

Derrom, Andrew Theodore. *As I Saw It: History in the Making*. New York: Carlton Press, 1974.

Drashevska, Lubov. *The Geology of Paterson, New Jersey With a Field Guide*. Paterson, NJ: Paterson Museum, 1976.

Four Chapters of Paterson History. Paterson, NJ: Lont & Overcamp, 1919.

Genovese, Peter. *New Jersey Curiosities*. Guilford, CT: Globe Pequot Press, 2007.

Goldman, Emma. "Observations and Comments: On the Paterson Silk Strike." *Mother Earth* 8, no. 1 (March 1913).

Green, Martin Burgess. *New York 1913: The Armory Show and the Paterson Strike Pageant*. New York: Scribner, 1988.

Gutman, Herbert G. *The Reality of the Rags-to-Riches "Myths": The Case of the Paterson, New Jersey Locomotive, Iron and Machinery Manufacturers, 1830–1980*. Oxford, New York: Basil Blackwell, 1977.

Hamilton, Alexander. *Alexander Hamilton: A Biography in His Own Words*. Edited by Mary-Jo Kline. New York: Harper & Row, 1973.

———. *Report on Manufactures*. 1791.

Historic American Engineering Record (HAER). *Great Falls SUM Survey Report on the First Summers Work in Cooperation with the Great Falls Development Corporation*. Paterson, NJ: 1973.

Johnson, Paul E. *Sam Patch: The Famous Jumper*. New York: Hill & Wang, 2003.

Kennedy, J. *The Falls of Passaic—A Visit from New York to Paterson*. New York: B. Waugh & T. Mason, 1833.

Longwell, Charles Pitman. *Historic Totowa Falls*. Paterson, NJ: self-published, 1942.

———. *A Little Story of Old Paterson as Told by an Old Man*. N.p.: self-published, 1901.

Murphy, J. Palmer, and Margaret Murphy. *Paterson & Passaic County: An Illustrated History*. Northridge, CA: Windsor Publications, 1987.

Nelson, William. *Historical Sketch of Passaic County, New Jersey, Especially of the First Settlements and Settlers*. Paterson, NJ: Chiswell & Wurts, 1877.

———. *History of Paterson and Its Environs (the Silk City)*. New York: Lewis Historical, 1920.

———. *History of the City of Paterson and the County of Passaic*. Paterson, NJ: Press Printing & Publishing, 1901.

Norwood, Christopher. *About Paterson: The Making and Unmaking of an American City*. New York: Saturday Review Press, 1974.

Pachman, Morris. "History of Dolphin Jute Mills, Paterson, New Jersey." MA diss., New York University, 1954.

Paterson Strike Pageant Program. Edited by Frederick Boyd. New York: self-published, 1919. Reprinted in *Rebel Voices: An IWW Anthology*, edited by Joyce L. Kornbluh. Ann Arbor: University of Michigan Press, 1964, 210–14.

Rea, Ann Webb. *The Ivanhoe Papermill*. Paterson, NJ: self-published, 1974.

Read, Philip M. *Paterson*. Charleston, SC: Arcadia Publishing, 2003.

Shriner, Charles A. *New Jersey: Its Advantages for Manufacturing*. Paterson, NJ: Press Printing and Publishing, 1800.

———. *Paterson, New Jersey, A Modern Day City*. Paterson, NJ: Press Printing and Publishing, 1910.

Society for Establishing Useful Manufacturers. Meeting Minutes. Available at the Free Public Library, Paterson, NJ.

Stansfield, Charles A., Jr. *A Geography of New Jersey: The City in the Garden*. New Brunswick, NJ: Rutgers University Press, 1998.

Stein, Beth. "The Immigrant and the Paterson Silk Strike of 1913." Paper presented at Seton Hall University, South Orange, New Jersey, August 6, 1973.

Tice, George A. *Paterson*. New Brunswick, NJ: Rutgers University Press, 1972.

Tripp, Anne H. *The IWW and the Paterson Silk Strike of 1913*. Champaign: University of Illinois Press, 1987.

Trumbull, Levi. *A History of Industrial Paterson*. Paterson, NJ: Carleton M. Herrick, 1882.

Wallerstein, Jane. *Voices from the Paterson Silk Mills*. Charleston, SC: Arcadia Publishing, 2002.

White, John H., Jr. *A History of the American Locomotive: Its Development from 1830–1880*. Mineola, NY: Dover Publications, January 1980.

Widmer, Kemble. *The Geology and Geography of New Jersey*. New Jersey Tercentenary Commission: Van Nostrand, 1964.

Williams, William Carlos. *Paterson*. New York: New Directions Paperbook, 1963.

Wyckoff, William. *American Silk Manufacture*. New York: self-published, 1887.

Index

About the Author

Marcia Dente was a lifelong resident of the city of Paterson until 2007, when she moved to South Jersey. For more than fifty years, she lived on Front Street, approximately two blocks from the Great Falls, and watched over that span of time as the area surrounding the Falls developed into a national natural resource, a historic landmark, a state park and, in November 2011, a national historical park.

Long before the Falls had any recognition, she played in the area behind the Falls and walked across the Wayne Avenue Bridge every day on her way to high school. She watched, painted and photographed the Falls as the water cascaded over the rocks from the Passaic River.

Dente retired in 2005 after forty-five years of service to the City of Paterson. Seventeen of those years were spent working with the Paterson Free Public Library, and the remaining years were spent working with the Department of Public Works. During the years with the library, she was the staff artist in the publicity and art department and worked on many projects, which included poetry contests, publishing poetry anthologies and judging art contests. She primarily assisted Mary Ellen Kramer, wife of former mayor Lawrence "Pat" Kramer, in the creation of the Great Falls Festival in 1971, which continues to this day.

Little did she know that when she was transferred to the Department of Public Works in 1977, she would continue to be involved in the planning stages of the festivals, assisting the late Ruth Hirshberg, director of special events; Paul Vetreno, superintendent of Parks and Shade Trees in the 1970s; and, later on, with other directors of the Department of Public Works and city officials.